GREENWASH
The Reality Behind
Corporate Environmentalism
by Jed Greer & Kenny Bruno

About the authors:

Jed Greer is a researcher on transnational corporations and hazardous technologies, formerly with Greenpeace International. His articles have appeared in *The Ecologist*, *Third World Resurgence*, and *Multinational Monitor* magazines, and he is the co-author (with Kavaljit Singh) of *TNCs and India: An Activist's Guide to Research and Campaign on Transnational Corporations* (Public Interest Research Group, Delhi, 1996).

Kenny Bruno has worked for Greenpeace since 1985 on a variety of toxics issues, and is currently a technology transfer specialist with Greenpeace International. He has written articles for *Third World Resurgence* and the *Multinational Monitor*, and his publications include *Screening Foreign Investments* (Third World Network, 1994), which he compiled, and the *Inventory of Toxic Technologies* (Greenpeace, 1994), which he edited.

GREENWASH
The Reality Behind
Corporate Environmentalism
by Jed Greer & Kenny Bruno

TWN
Third World Network
Penang, Malaysia

The Apex Press
New York, USA

GREENWASH
The Reality Behind Corporate Environmentalism
is published by:

Third World Network,
228 Macalister Road,
10400 Penang,
MALAYSIA

and

The Apex Press
777 United Nations Plaza, Suite 3C
New York, NY 10017,
USA

cover illustration by Paul Normandia

Printed by Jutaprint,
2 Solok Sungei Pinang 3,
Sungei Pinang, 11600 Penang,
Malaysia.

ISBN: 0-945-257-77-5

To NCG & AWG

ACKNOWLEDGEMENTS

This book is an expanded version of the 1992 report "Greenpeace Book on Greenwash," released at the Earth Summit in Rio de Janeiro.

We wish to extend sincere appreciation to our many Greenpeace colleagues around the world who supplied information and editorial assistance. We are grateful to Jim Vallette as well as Curt and Leslie Greer and their family for various help and support.

Many thanks are also due to the staff of Third World Network who helped produce this book: Martin Khor, Lim Beng Tuan, Linda Ooi, and Lim Jee Yuan.

CONTENTS

INTRODUCTION

"Industry will have the primary role in making [sustainable development] work. We are the experts at development."
– *Edgar S. Woolard, Jr.,*
DuPont Chairman[1]

"The leadership of the North in the generation of environmentally unsound technologies does not automatically translate into a leadership to generate environmentally sound technologies."
– *Vandana Shiva, Director, Research Foundation for*
Science, Technology, and Natural Resource Policy, India [2]

A corporate leader in ozone destruction takes credit for being a leader in ozone protection. A giant oil transnational embraces the "precautionary approach" to global warming. A major agrochemical manufacturer trades in a pesticide so hazardous it has been banned in many countries, while implying that it is helping to feed the hungry. A petrochemical firm uses the waste from one polluting process as raw material for another, and boasts that this is an important recycling initiative. A logging company cuts timber from natural rainforest, replaces it with plantations of a single exotic species, and calls the project "sustainable forest development." And these corporations, with the help of their business associations and public relations firms, help set the agenda for global negotiations on the crises of environment and development.

Welcome to the world of "greenwash," where transnational corporations (TNCs), which increasingly dominate the world economy, are preserving and expanding their markets by posing as friends of the environment and leaders in the struggle to eradicate poverty.

In the 1990s, greenwash has gone global as TNCs and their environmental public relations efforts gain unprecedented influence over world affairs. With the cooperation of governments and intergovernmental organizations, TNC greenwash has controlled environmental approaches in the United Nations system, at international and regional trade negotiations, at lending institutions, and at the 1992 United Nations Conference on the Environment and Development (UNCED or "Earth Summit") in Rio de Janeiro as well as its follow-up.

Through a sophisticated greenwash strategy, TNCs are working to manipulate the definition of environmentalism and of sustainable development, and to ensure that trade and environment agreements are shaped, if not dictated, by the corporate agenda. Giant corporations have made international political affairs a part of their strategy to convince the public that they have turned the corner into a new era of "green business."

This book provides evidence that despite the rhetoric, TNCs have not substantially changed their environmental behavior in this new era. We trace the phenomenon of greenwash, examine corporate "self-regulation," detail the activities of corporate lobbying groups in the UNCED process, and profile the words and deeds of 20 large corporations which proclaim their environmentalism.

A look at corporate behavior exposes the reality hidden under the green image being created by TNCs. The reality is that TNCs are not saviors of the environment or of the world's poor, but remain the primary creators and peddlers of dirty, dangerous, and unsustainable technologies. The claims of these companies must be scrutinized carefully and their activities and products regulated for the good of the planet and its people. This book starts such scrutiny by examining 20 greenwashing corporations headquartered in nine countries on four continents, with operations in almost every part of the world, and which profit from major dirty and destructive industrial sectors.

Before Greenwash: Typical Corporate Responses to Environmental Problems

The heavily publicized environmental orientation of TNCs is a relatively new phenomenon, but it should be understood in the context of traditional corporate responses to ecological and social damage. Often, corporations have simply denied problems, avoided responsibility, resisted controls, and engaged in "job blackmail."

For example, DuPont and other chlorofluorocarbon (CFC) manufacturers denied their role in ozone depletion for over 14 years after scientists first discovered a connection between their product and the destruction of ozone molecules. During the years that they fought to downplay the scientific findings, chlorine built up in the atmosphere and depletion of the ozone layer accelerated worldwide.[3]

Union Carbide, which ran the Bhopal pesticide plant, has attempted to avoid responsibility for the deaths and injuries from the worst corporate industrial accident in history. When Carbide agreed in 1989 to pay US$470 million in damages, the Indian government dropped all criminal charges against the company.[4] That settlement was finalized in 1991, but as of 1993 most surviving victims had received little or no compensation.[5] In 1991, criminal charges were reinstituted against the company and a number of its executives. In 1994, however, Carbide began to sell its assets in India and move out of the country, making it especially unlikely that the corporation or its officials will ever face the charges in court.[6] Similarly, Dow Chemical and Shell Oil engaged in legal maneuvers for years in a suit against them for manufacture and use of DBCP, a pesticide which allegedly caused sterility in Costa Rican farmworkers.[7]

As a group, TNCs and their political associations have resisted laws and proposals for controlling CFC production, hazardous waste disposal, international trade in wastes and banned products, taxes on hazardous substances, cutting of virgin forests, fuel and energy efficiency, and countless other issues, through lobbying, financial contributions to officials who will represent their views in government, and other similar methods.

When denial, avoidance, and resistance fail, corporations can resort to "job blackmail," that is, they threaten to move production elsewhere if measures they oppose are implemented. More and more, companies can and do move, often avoiding environmental regulations and employing cheaper labor in the process.

Of course, avoidance, resistance, and "job blackmail" are still common corporate reactions to environmental protection and worker safety initiatives. But by the late 1980s the sheer weight of evidence of the devastating ecological and social impacts of corporate operations was eroding the plausibility and effectiveness of simple denial.

Furthermore, in the 1970s and 1980s the growth of community-based movements against environmental degradation in many countries, the emergence of environmental platforms from Green and other political parties in Europe and elsewhere, and increasing media exposure of ecological problems all contributed to an unprecedented rise in environmental awareness. Public outcry against corporate activities increased as people traced the ozone hole, dying lakes and seas, disappearing forests, the changing global climate, and toxic waste dumps scattered around their communities back to decisions made by TNC managers.

Birth of Greenwash

By the late 1980s, this popular movement had gained sufficient strength to become a potential threat to the political power and financial health of transnational corporations. TNCs could no longer deny their role in environmental degradation. Instead, they embraced the environment as their cause and co-opted its terminology in advertisements and corporate policies. While little changed in practice, the greenwash counterstrategy was born.

In the early 1990s – and just in time for UNCED – corporations had expanded the greenwash concept and were presenting themselves as proponents of "sustainable development." This position was stated clearly by the premier TNC association which lobbied at UNCED, the Business Council for Sustainable Development: economic growth through deregulated free trade and "equitable access to markets for all" are "an essential prerequisite both for sustainable development and for the continuing prosperity of the more industrialized nations."[8] They held up free trade as a panacea, arguing that it will produce enough growth to end poverty, and after that generate resources for environmental protection.

The unrestricted free trade and investment-based growth beloved by TNCs is the same kind of development which has led to overexploitation of land and natural resources, air, water, and soil pollution, ozone depletion, global warming, and toxic waste generation. As economists Herman Daly and Robert Goodland observe: "The dream that growth will raise world wages to the current rich country level, and that all can consume resources at the U.S. per capita rate, is in total conflict with ecological limits that are already stressed beyond sustainability."[9] The wealth created by such growth has never caught up to the social and environmental problems it creates. The addition of the word "sustainable" – to be defined by the corporation itself – is primarily a cosmetic change to this traditional approach.

The Role of TNCs in Environmental Destruction

TNC operations routinely expose workers and communities to an array of health and safety dangers. All too often these operations erupt into disasters. Many large-scale accidents occur outside the corporation's home country, and some involve double standards in health and safety as well as environmental regulations and enforcement. A few of the most infamous examples include:

Seveso, Italy – A subsidiary of the Swiss-based pharmaceutical company Hoffman-LaRoche set up a trichlorophenol plant for which permits could not be obtained in Switzerland; a 1976 explosion there exposed 156 workers and 37,000 residents to a toxic dioxin cloud. [10]

Bhopal, India – A major gas release in 1984 at the pesticide plant of US-based Union Carbide's Indian affiliate led to the death of thousands of people and injuries to hundreds of thousands more.[11]

Basel, Switzerland – A huge spill of disulfoton and other chemicals into the Rhine River in 1986 killed fish, wildlife, and plants for hundreds of miles. Swiss-based Sandoz, responsible for the majority of the spill, subsequently moved its disulfoton warehouse to Resende, Brazil.[12]

To regard such tragedies only as "accidents," however, distracts attention from the inherent dangers of the industrial development strategy in which TNCs play so important a part. For example:

Fossil Fuels – TNC activities generate more than half of the greenhouse gases emitted by the industrial sectors with the greatest impact on global warming. TNCs control 50 per cent of all oil extraction and refining, and a similar proportion of the extraction, refining, and marketing of gas and coal.[13]

CFCs – TNCs have virtually exclusive control of the production and use of ozone-destroying chlorofluorocarbons (CFCs) and related compounds.[14]

Mining – In minerals extraction, TNCs still dominate key industries. In aluminum, just six companies control 63 per cent of the mine capacity, 66 per cent of the refining capacity, and 54 per cent of the smelting capacity.[15]

Agriculture – TNCs control 80 per cent of land worldwide which is cultivated for export-oriented crops, often displacing local food crop production. Twenty TNCs account for about 90 per cent of pesticide sales, and control much of the world's genetic seed stocks. TNCs finance the bulk of biotechnology research worldwide.[16]

Toxic Chemicals – TNCs manufacture most of the world's chlorine – the basis for some of the most toxic, persistent, and bioaccumulative synthetic chemicals known including PCBs, DDT, dioxins and furans, chlorinated solvents, and thousands of other organochlorine compounds. These chemicals' impacts on health include: immune suppression; birth defects; cancer; reproductive, developmental, and neurological harm; and damage to the liver and other organs.[17]

Trade in Hazards – TNCs as a group lead in the export and import of products and technologies that have been controlled or banned in some countries for health and safety reasons. For example, 25 per cent of total pesticide exports by TNCs from the US in the late 1980s were chemicals that were banned, unregistered, canceled, or withdrawn in the US itself.[18] A single TNC controls nearly the entire world market in tetraethyl lead gasoline additive, a product virtually eliminated in the US, Canada, Japan, and Australia, and being phased out in Western Europe due to its well-known contribution to environmental lead contamination and childhood lead poisoning.[19] And a handful of Northern companies are responsible for the nuclear technology now found at plants in South America and Asia.

The Power Behind Greenwash: An Overview of TNCs in the World Economy at the End of the 20th Century

In the pursuit of industrial development around the globe, TNCs are joined by a host of other allied institutions: multilateral development banks, bilateral aid agencies, free trade organizations, and national governments. Yet among all these important players, TNCs have emerged as the most significant actors in the world's economy.

There has been a proliferation of transnationals over the past quarter century. In 1970, there were some 7,000 parent TNCs. Today, that number has jumped to 37,000, 90 per cent of them based in the industrialized world, which control over 200,000 foreign affiliates. These subsidiaries generated global sales of US$4.8 trillion in 1991, rising to US$5.8 trillion in 1992. TNCs control an estimated 70 per cent of global trade and hold 90 per cent of all technology and product patents worldwide.

The large number of TNCs can be somewhat misleading, because the wealth of transnationals is concentrated among the top 100 firms (16 of which are profiled in this book) that account for about one-third of the combined outward foreign direct investment (FDI) of their countries of origin. In 1992, the largest 100 TNCs had US$3.4 trillion in global assets, of which approximately US$1.3 trillion was held outside their home countries.

Comparison of TNCs' Foreign Affiliates' Sales with World Exports		
	1991	1992
World Sales of TNCs' Foreign Affiliates	4.8	5.8
World Exports of Goods & Services	4.5	4.7
(figures in US$ trillions) **Source**: *United Nations*, 1994		

Beginning in the mid-1980s, a large rise of TNC-led foreign direct investment has occurred. Between 1988 and 1993, worldwide FDI stock – a measure of the productive capacity of TNCs outside their home countries – grew from US$1.1 to US$2.1 trillion in estimated book value. The annual rate of growth of FDI stock since 1991 has been greater than that of worldwide exports of goods and services, worldwide gross domestic product, gross domestic investment, and receipts for technology payments. The share of FDI inflows in gross fixed capital formation more than doubled between 1986 and 1992. FDI outflows from all countries was US$193 billion in 1993, increasing to US$204 billion in 1994.[20]

Foreign Direct Investment (FDI) and Portfolio Investment (PI) Inflows to Africa, Asia, and Latin America, 1986-1992

Region	1986-90 annual average	1991	1992
Africa			
FDI	2.6	2.5	2.9
PI	-	-	0.1
Latin America & the Caribbean			
FDI	6.4	12.4	14.5
PI	-	8.9	5.5
Western Asia			
FDI	0.3	0.5	0.3
PI	-	-	-
South, East, & Southeast Asia			
FDI	7.5	14.5	21.1
PI	2.2	4.3	8.1
Total	19	43.1	52.5

(figures in US$ billions)
Source: *United Nations*, 1994

FDI Outflows from the Five Major Home Countries, 1982-1994

Country	1982-86	1987-91	1990	1991	1992	1993	1994
		annual average					
France	3	20	35	24	31	21	25
Germany	6	18	29	23	16	15	15
Japan	7	35	48	31	17	14	18
United Kingdom	10	28	19	16	19	26	27
USA	11	28	30	31	41	58	56
Total	37	129	161	125	124	134	141
Total All Countries	55	192	239	191	194	193	204

(figures in US$ billions)
Source: *United Nations*, 1994

Transnationals participate in a growing number of business collabo-
rations such as joint ventures, and have been heavily involved in interna-
tionalizing national financial markets.[21] TNCs are also benefiting from
investment liberalization and privatization policies many governments
have introduced in recent years, as well as from political changes which
have opened new opportunities in Central and Eastern Europe and the
former Soviet Union.

Moreover, there has been a great increase in TNC investment in the
less-industrialized world. FDI into less-industrialized nations rose five-
fold between 1986 and 1993. In 1992, foreign investment flows into those
countries was US$50 billion; in 1993 the figure had jumped to US$80
billion. In 1992-93, less-industrialized countries accounted for between
one-third and two-fifths of global FDI inflows – more than at any time
since 1970. These flows have not been evenly distributed, however, with
just ten host recipients accounting for up to 80 per cent of all FDI to the
less-industrialized world.[22]

Overall since the mid-1980s, financial flows to less-industrialized
countries from corporate investment and private bank loans to support
such investment have grown far more dramatically than those from
national development aid or multilateral bank lending. To be sure, the
estimated US$40 billion which multilateral development banks annually
lend, and of which TNCs are key beneficiaries, still have an enormous
impact on the lives and livelihoods of people across the globe as well as on
the environment. Often, that impact has been – and continues to be –
extremely destructive. This is especially true in the case of the World Bank,
whose 50th anniversary in 1994-95 was celebrated by hundreds of organizations around the world in the "50 Years Is Enough" campaign, which aims
at restructuring the Bank or eliminating it entirely.

To some degree, however, private foreign investment and loans to
less-industrialized nations are now replacing the historical function of
institutions such as the World Bank. As one commentator explains in a
1994 article in the British magazine *The Economist*:

"The traditional project-financing role of the World Bank is rapidly
becoming passe, as the Bank's own 'World Development Report'
indicates this year. It rightly points out that world capital markets are

mobilising project finance of a kind that was unthinkable a few years ago. The markets are channelling tens of billions into roads, ports, telecoms and power generators – and thereby supplanting the long-standing role of the Bank itself. Many countries are turning away from the Bank's project loans because private ones are cheaper."[23]

The Bank is trying to adapt to these circumstances, in part by increasing its involvement with private-sector projects. For instance, it is boosting loans made by its subsidiary, the International Finance Corporation (IFC), which lends to private companies, frequently for joint ventures with TNCs.

Sovereignty Eroded

There is no gainsaying the augmented prominence and influence of transnational corporations and their foreign investment in global economic affairs. Such power has in turn led to TNCs' increased political sway. Commonly, say Richard J. Barnet and John Cavanagh in their 1994 book *Global Dreams: Imperial Corporations and The New World Order*, top corporate executives describe their TNCs as "stateless," willing and increasingly able to move their operations in search of lower costs.[24] Greater geographic flexibility has not only intensified rivalry between corporations. It has also sparked competitive efforts between nations to attract TNC investment, with inducements being low wages, weak health and safety laws, and lax environmental standards. "The decline of the political power and technical means of national governments to regulate the behavior of global corporations operating on their territories," write Barnet and Cavanagh, "has helped to bring about an ideological shift that makes a virtue out of this reality."[25]

A crucial aspect of this ideological shift has been the push to reduce barriers to trade and investment capital flows. Transnationals have always had a big stake in trade and foreign investment rules negotiations, and have worked hard in the last decade to shape to their liking Europe's Single Market agreement, the North American Free Trade Agreement (NAFTA), and the Uruguay Round of the General Agreement on Tariffs and Trade (GATT). [26]

For TNCs, so-called free trade lessens governmental restrictions on their movement and ability to maximize returns. "The deregulation of trade aims to erase national boundaries insofar as these affect economic life," Herman Daly and Robert Goodland have noted. "The policy-making strength of the nation is thereby weakened, and the relative power of TNCs is increased."[27] "The more free trade, the better for us," Asea Brown Boveri's Chief Executive Percy Barnevik acknowledges. "We can optimize our production globally."[28]

In the 1990s, environmentalists finally woke up to the frightening ecological implications of deregulated free trade and have joined workers and farmers in organizing considerable opposition. In response, free trade supporters now routinely point to the environmental safeguards free trade agreements will include. Even the Bush Administration felt compelled to insist that NAFTA was the "greenest" trade agreement ever.[29]

Yet analyses of NAFTA and GATT show how these agreements offer innumerable opportunities for undermining national attempts to protect the environment as well as tending to lower wages and safety standards in all countries.[30] The US-Canada Free Trade Agreement has already demonstrated these tendencies clearly. Under this agreement, the US has attempted to require Canada to abandon Pacific salmon protection, to lower pesticide standards, and to end reforestation subsidies. And Canada has challenged the US ban on asbestos and recycled newsprint requirement.[31]

UNCED's Secretary-General Maurice Strong, in his opening statement at one of the Conference's Preparatory Committee negotiations, called for UNCED to be made consistent with GATT. UNCED itself may have been a preview to the subservience of environmental agreements to the priorities of deregulated free trade. As TNCs lobbied in the Uruguay Round of GATT and other free trade negotiations to open more markets and eliminate regulations, they simultaneously joined with the US, the European Union, and Japan to make UNCED consistent with GATT – thus forcing an undesirable marriage of the concepts of unrestricted free trade and sustainable development, with free trade as the dominant partner.

Rules established in the GATT's Uruguay Round regarding trade-related intellectual property rights (TRIPs) and trade-related investment measures (TRIMs) will be of particular benefit to corporations. The first gives TNCs greater capacity to privatize and patent life itself, including plant and other genetic resources of less-industrialized nations and peoples. TRIMs render illegal certain measures which countries – especially Southern nations – have employed to encourage TNCs to establish linkages with domestic firms. TRIPs, TRIMs, and other GATT rules fall under the authority of the World Trade Organization (WTO), a new supranational body which works with the World Bank and other financial institutions to manage global economic policy to serve transnational corporate interests.[32]

With powerful international momentum for deregulated trade and intensified competition between nations for foreign investment, countries are in a manifestly weaker position to control TNC activities than they were two decades ago. "[T]he bargaining power of governments vis-a-vis foreign capital has declined," assert Barnet and Cavanagh, and with it political leaders' ability to address "such problems as unemployment, ecological deterioration, and the corrosive effects of chronic poverty."[33] "The retreat of government in many places around the world," they note further, "has left a power vacuum that corporations have rushed to fill in."[34]

One consequence of this "retreat" is that although TNCs are collectively the world's most powerful economic force, no intergovernmental organization is charged with regulating their impact on the environment and development crises faced by the nations of the world. Opposition to international regulations on TNCs has come not only from industry but also from governments, say Riva Krut and Harris Gleckman, formerly of the United Nations Centre on Transnational Corporations. "The tragedy," they write, "is that it is in the international arena that issues of global consequence are being addressed, and national resources could well be used here. Instead, while international economic ties and the powers of TNCs have grown and developed...the strength of complimentary institutions, such as the United Nations, has diminished."[35]

United Nations efforts to monitor TNCs have indeed been weakened. Under a restructuring ordered by United Nations Secretary-General Boutros-Ghali, the UN's economic, social, and environmental programs have been deemphasized in favor of security and policing functions. In 1992, just before the Earth Summit, the UN Centre on Transnational Corporations (CTC) lost its independent status. In 1993, the CTC was essentially dismantled, its staff distributed among other UN agencies, and a new Division on Transnational Corporations and Investment emerged, whose role is to promote foreign direct investment. A 17-year attempt to negotiate an overarching Code of Conduct on TNC behavior was abandoned. These changes are consistent with recommendations the ultra-conservative US Heritage Foundation made in 1991, and represent a victory for the TNCs in their fight to stay off the United Nations agenda.[36]

Greenwash Goes Global: The Corporate Hijack of UNCED

In the age of globalization, it is not enough for individual TNCs to defend themselves against accusations of environmental crimes and to improve their environmental image. Nor are they content with lobbying on national legislation. Individually and collectively, TNCs are attempting to expand their influence in international affairs, especially trade, environment, and development agreements. UNCED is perhaps the best example to date; corporate influence on the Earth Summit undermined parts of Agenda 21 (the 800-page document intended to provide an action plan for future work on sustainable development), rendered the UN Framework Convention on Climate Change toothless, and weakened the Convention on Biodiversity, which was nonetheless rejected by the United States under the Bush Administration.

Proposals to regulate, or even monitor, the practices of large corporations were mostly removed from UNCED documents. In fact, the treatment of TNCs at the Earth Summit was based on the assumption of industry itself – that Northern-based corporations have the know-how and the capacity to spread environmentally sound, sustainable technologies throughout the world. The Earth Summit itself was greenwash on a grand scale because it gave the false impression that important, positive change was occurring and failed to alert the world to the root causes of environment and development problems.

Agenda 21

In March 1992, during the fourth and final Preparatory Committee (PrepCom) meeting for the Earth Summit, the International Chamber of Commerce (ICC) and its members pressed hard to keep any language calling for regulation of the TNCs out of Agenda 21 and the Earth Charter, the two key documents produced at UNCED. Aided by the US government and others, the TNCs even went so far as to advocate successfully striking the words "transnational corporation" from many of the texts. In one case, the ICC took its fight to Stockholm, where it lobbied the Swedish government to withdraw a proposal it had made calling for TNCs to internalize environmental costs in their accounting and reporting processes, and replace it with "inviting them [TNCs] to participate in examining the implications for internalizing environmental costs."[37]

More recently, regulation of TNCs was also conspicuously absent from the agenda of the 1995 World Summit for Social Development in Copenhagen. Summit Preparatory Committees had drafted language calling for a voluntary code of conduct for transnationals. But by the time of the conference, the idea had been dropped and there is virtually no mention of TNCs in Social Summit documents.[38]

Another area of Agenda 21 which bore the imprint of corporate lobbying is the chapter on technology transfer. Arguing that they have the technology that will allow for sustainable development in the South, TNCs worked throughout the UNCED negotiations for a definition of technology transfer which they call "technology cooperation." "Technology cooperation" is based on US-style patent, copyright, and intellectual property rights laws which permit corporations to keep tight control of materials and information as trade secrets, even in cases where the information for developing new technology – such as seed stock or genetic material – comes from sources in the South. Such "technology cooperation" will ensure that the benefits and profits of new technology stay with the corporations, not the country of origin, or the country where it is being developed or tested. In addition, "technology cooperation" can serve to make Southern countries permanently dependent on capital-intensive imported equipment, spare parts, and skills.[39]

Agenda 21 defines environmentally sound technologies as those "that protect the environment, are less polluting, use all resources in a more sustainable manner, recycle more of their wastes **in a more acceptable manner than the technologies for which they were substitutes**" (emphasis added).[40] Under this definition, all environmental soundness is relative. If a TNC imports a technology from Sweden to Poland, for example, it would be considered environmentally sound if it were cleaner than the previous technology in Poland, even if it were considered dirty by Swedish standards. In this hypothetical scenario, the TNC gets credit for transferring environmentally sound technology, lending institutions may support the project, but the urgent local and global need to leapfrog current dirty technologies is ignored.

Climate Change Convention

Three hundred scientists from 40 countries made up the Intergovernmental Panel On Climate Change (IPCC) which produced the 1990 report "Climate Change: The IPCC Scientific Assessment." This document states explicitly that the only hope for avoiding unprecedented and ecologically disastrous global warming is to make deep cuts in carbon dioxide emissions. The Climate Change Convention, which did not obligate any country to reduce such emissions, therefore ranks as one of UNCED's most devastating failures.

In part, the weakness of the Convention represented the payoff of the work of the Global Climate Coalition, an industry lobby which was active throughout the two-year negotiating process. At the last session of the pre-conference negotiations, the Coalition passed out fliers which claimed, contrary to all the evidence amassed by the IPCC, that no environmental benefit will be achieved by stabilizing carbon dioxide emissions.

Shell and DuPont, both members of the Coalition, profess the precautionary approach to global warming. But a look underneath the fancy green language of "eco-efficiency," "no-regrets," and "precautionary principle," reveals a business-as-usual approach within the oil industry. After a press conference in Rio, Gabriele Cagliari, former Chairman of the Italian oil giant ENI, was asked if the world can burn all the oil on the planet and call it sustainable. "Yes," Cagliari answered.[41]

Source: Picture of cover of BCSD's book *Changing Course*, The MIT Press, 1992.

Biodiversity Convention

The biggest scandal of the Earth Summit was the US President Bush's decision not sign the Biodiversity Convention, and his rejection of the US delegation leader's last-minute recommendation that the US allow Brazil to try to renegotiate the treaty so the US could sign. Subtle greenwash was not necessary to convey his position. Pointing out that he was "the president of the United States, not the president of the world," Bush himself explained that "in biodiversity it is important to protect our rights, our business rights." [42] Nor did the Clinton Administration's decision to sign the Convention mean a victory for biodiversity over "business rights." The US signature came with an "interpretive statement" to the Convention which emphasizes biotechnology companies' interest in US-style intellectual property protection. [43]

Greenwash Elite – The Business Council For Sustainable Development

While the ICC, Global Climate Coalition, International Council on Metals and the Environment, and their members were doing the dirty work of lobbying at UNCED and back home in capital cities, another business organization was undertaking an enormous public relations drive. The Business Council for Sustainable Development (BCSD) – a grouping of some 48 chief executives or chairmen from industrial sectors including energy, chemicals, forestry, pesticides, transportation, finance, and communications – attempted to take over the UNCED stage to claim TNCs had voluntarily turned the corner onto a new path of sustainability and are leaders of sustainable development.

The BCSD was established late in 1990 at the specific request of UNCED Secretary-General and millionaire businessman Maurice Strong. Strong himself gave meticulous assistance to the BCSD in the presentation of its proposals and helped the Council edit *Changing Course – A Global Business Perspective on Development and the Environment* – the centerpiece of industry's submissions at Rio. He later paid public tribute to the "extremely important contribution" which the Council made to the Earth Summit.

Until 1994, the BCSD was chaired by Swiss billionaire Stephan Schmidheiny, whom Strong appointed as his Principal Advisor for Business and Industry. Among his corporate endeavors, Schmidheiny has served on the boards of directors of both Asea Brown Boveri, which is involved in the marketing and building of nuclear reactors (and is profiled in this book), and Nestle, which has been the target of consumer boycotts for its aggressive marketing of infant formula as a substitute for breast milk to women in less-industrialized countries. Schmidheiny's family business interests have included asbestos production in Brazil. And while his image makers promoted his small crafts foundation in Latin America, this venture is microscopic compared to Schmidheiny's 30 per cent interest in Chile's largest steelmaking concern.[44]

In the post-UNCED period, the BCSD set up national and regional affiliates in Latin America and Asia as well as a binational Gulf of Mexico BCSD, housed in the state government offices of Texas. In late 1994, the BCSD agreed to merge with another industry group, the World Industry Council for the Environment (WICE), which the International Chamber of Commerce initiated in 1993. WICE's 85 members include Royal Dutch/Shell, ICI, Ciba, Sandoz, Rhone-Poulenc, Mitsubishi, and Mobil. The new body is called the World Business Council for Sustainable Development (WBCSD).[45]

The Greenwash Professionals – Burson-Marsteller

One of the BCSD's primary targets at the Earth Summit was the hearts and minds of the global public. To present its case at the Rio Summit, the BCSD hired public relations giant Burson-Marsteller (B-M). B-M has had considerable experience in helping TNCs with PR problems:

* In the 1970s, when Babcock & Wilcox's global sales suffered after the nuclear reactor it built failed at Three Mile Island, Burson-Marsteller was there to assist its client.[46]

* When A.H. Robins could no longer handle the international public relations woes resulting from the problems with its Dalkon Shield contraceptive device, it called on B-M.

- In the aftermath of Union Carbide's Bhopal gas disaster, Burson-Marsteller advised the company.

At other times in B-M's 40-year history, governments have turned to the firm for "issues management." During the reign of Romania's Nicolae Ceausescu, for example, Burson-Marsteller was hired to promote the country as a good place to do business.[47] When the former military dictatorship of Argentina was having difficulty attracting international investment, the ruling military junta hired B-M to "improve the international image" of the country over a period during which some 35,000 people were "disappeared."[48] More recently, B-M has served as the lobbyist for the Mexican government, promoting the environmentally questionable free trade agreement between Mexico, the US, and Canada.

B-M takes pride in the professional nature of its greenwash activities:

> "Often corporations face long-term issue challenges which arise from activist concerns (e.g. South Africa, infant formula) or controversies regarding product hazards... Burson-Marsteller issue specialists have years of experience helping clients to manage such issues. They have gained insight into the key activists groups (religious, consumer, ethnic, environmental) and the tactics and strategies of those who tend to generate and sustain issues. Our counselors around the world have helped clients counteract [them]."[49]

Burson-Marsteller's services don't come cheap, and with B-M the BCSD joined a corporate tradition of spending large resources not on actual environmental change, but on creating a "green image" for the client. Of course, Burson-Marsteller is not the only PR firm helping business paint itself green. Greenwash around the world bears the mark of professional, multi-million dollar public relations campaigns.

The Greenwashing of Corporate Culture

As part of the greenwash counterstrategy, corporations have notified the public that there has been a profound change in corporate culture. Some common manifestations of this new concern for environmental image and performance are:

- Corporate restructuring to include environmental issues, e.g., environmental officers at high levels, or new environmental departments within a corporation.

- Corporate environmental programs like waste minimization, waste reduction, and product stewardship.

- Responses to public concern about the environment; sometimes, these responses take place even when not required by law.

- Environmental themes in advertising and public relations.

- Voluntary environmental policies, codes of conduct, and guiding principles.

With the creation of such programs, we are asked to believe that corporations are now something fundamentally different than what they were before. But the addition of an "environmental department" does not change the raison d'etre of a corporation. It is critical that citizen activists and governments look under the surface of such announcements and be aware of the overall context in which they exist.

Certain basic characteristics of corporate culture have not changed. For example, in overseas operations especially, the assertion is often made that the mere presence of the corporation, its products, technologies, jobs, and culture are inherently beneficial to the host population. As an international waste trader, Arnold Andreas Kuenzler, said about a planned hazardous waste landfill in Angola, "If it's good enough for the Swiss, it's good enough for the blacks." [50] In many cases, corporations further imply that the dirty industries they bring will be the primary method whereby Southern countries can gain enough wealth to have the "luxury" of a clean environment.

Another fundamental dynamic is that through aggressive marketing or interlocking relationships with "customers," manufacturers help create "demand." Corporations then proceed to abdicate responsibility for problems created by their products by passing responsibility to the users of those products. Responsibility is passed along by intermediate users (such as the automobile industry in the case of CFCs) until it reaches the individual consumer. In the end, individuals are held responsible for production and marketing decisions made by giant corporations. This has even been the excuse for marketing products in the South which have been banned or restricted in the North such as lead gasoline additive and some pesticides.

To understand why corporate culture is so impervious to change, it is necessary to bear in mind the brutally competitive global economic atmosphere – between countries as well as companies – which TNCs are both responding to and fueling. Richard J. Barnet and John Cavanagh make the obvious but essential point that corporations are not chartered to solve social or environmental ills, but "are in business to make products and sell services anywhere they can to make money."[51]

In their advertisements and slogans, however, TNCs are increasingly trying to give the impression that they are in business to help people and to solve environmental problems. DuPont's "Better Things For Better Living," and "Dow Lets You Do Great Things," are examples. People watching television or reading magazines with greenwash may be seduced into forgetting that the fundamental drive of corporate culture is not improving their lives. Rather, the emphasis is on achieving cost advantage: minimizing expenses and maximizing revenue. In the culture in which business executives operate, the profit motive and pressure to raise shareholder returns remain the most influential determinants of corporate behavior and the key criteria for judging corporate performance.

Thus, despite their stated commitment to environmentalism, TNCs typically continue to justify their current activities, and new investments, with a cost-benefit assumption which fails to include the vast majority of environmental costs. Measuring only direct costs and short-term profits, corporations may tout the benefits of jobs and products created, but externalize costs of pollution, waste, and long-term damage to people and the environment.

Inside the World of Greenwash: Corporate Codes of Conduct

Many TNCs have adopted "corporate codes of conduct," "guiding principles," and other voluntary environmental policies as part of their response to ecological problems. Because these codes are offered as evidence that industry is taking its responsibilities seriously and is prepared to respond to citizen demands, they must be examined closely. Some companies imply that voluntary adherence to a code can replace regulation and monitoring of industry. Rhone-Poulenc has put this belief into practice. Its Chairman Jean-Rene Fourtou, who formed a 25-company association called Enterprises pour l'Environnement, credits the group with success in heading off new regulations in favor of voluntarist initiatives.[52]

Skeptical observers of the codes typically say that while the rhetoric is pretty, practice hasn't yet changed enough. But a closer look at the actual texts of these codes reveals that even the skeptics are too trusting. The codes adopt environmental terminology, such as "environmentally sound" and "sustainable development," while subtly changing the meaning of key words to cover industry behavior. In the end, the new rhetoric and the acknowledgement of relatively superficial problems in voluntary codes divert attention from the fundamental environmental issue: products such as nuclear reactors and toxic chemicals form the lifeblood of many TNCs. The codes are themselves a form of greenwash.

Two of the major corporate codes are the chemical industry's Responsible Care Program and the International Chamber of Commerce Business Charter for Sustainable Development, also known as the Rotterdam Charter.

Responsible Care

Responsible Care is the name of the chemical industry's major program on environmental issues. It originated in Canada in 1984 and was adopted in the US as a direct response to the Valdez Principles, a code developed by environmental organizations for corporations following the 1989 Exxon Valdez oil spill. All members of the US Chemical Manufacturers

Association must sign on to Responsible Care as an obligation of membership. Chemical industry associations in Western Europe and, more recently, Latin America, are developing similar programs, and it is a point of pride among many chemical company executives.

The two aspects of Responsible Care consistently emphasized by the associations and individual members are a "commitment to continuous improvement" in health, safety, and the environment, and the "profound cultural change" it represents.[53] Responsible Care acknowledges that the chemical industry as a whole has not performed even to its own satisfaction and that change is needed. This gives citizens concerned about company practices some leverage, and is a welcome admission. But there are a number of serious problems with Responsible Care:

- The US Chemical Manufacturer Association's president has stated that Responsible Care will help citizens to track corporations, monitor their performance, and make suggestions. Toward that end, each company is supposed to conduct an annual self-evaluation. However, the evaluations are not available to the public. Without access to information – even that generated by the company itself – the public does NOT have the opportunity to track the corporation any more than it did before Responsible Care.

- Although one of the "Guiding Principles" of Responsible Care is to develop safe products, there are no criteria for what constitutes a safe product. Even the most dangerous products, such as banned pesticides and ozone-destroying chemicals, are judged "safe" by Responsible Care signers.

- Under the heading "Pollution Prevention Code," Responsible Care has two parts: waste and release reduction and waste management. While waste reduction is desirable, this blithe interpretation of "pollution prevention" makes the phrase meaningless. Waste reduction and management are often forms of end-of-pipe pollution control measures, not preventative measures. Pollution prevention should refer to the avoidance of toxic chemical production, use, and disposal in the first place. The text takes waste practices which are responsible for much of the pollution spread by TNCs and legitimizes them as "prevention."

- Responsible Care emphasizes "environmental performance," suggesting that the only thing wrong in the chemical industry is that there are too many "incidents." While a reduction in accidents and spills is vital, it is notable that Responsible Care does not acknowledge the inherent toxicity of many chemical company products and routine emissions. Thus, a corporation which increases production of an unnecessary and toxic product can claim to have improved "environmental performance" if it has had fewer accidents in the manufacturing process.

- Responsible Care does not apply to foreign subsidiaries of member companies. Company evaluations do not include overseas operations, and overseas environmental policies are not addressed. Other business charters and some companies are more comprehensive than Responsible Care in this area.

- Responsible Care has failed to take root in much of the chemical industry. In a 1992 survey of US Responsible Care member companies, the US Public Interest Research Group found that 42 per cent of the companies were unreachable by phone, and 27 per cent of those contacted refused or were unable to answer any questions about chemical use, storage, shipments, and related matters. Only ten per cent of the 192 facilities surveyed answered all the questions asked by the group.[54] The Chemical Manufacturer's Association, which organizes Responsible Care in the US, found in its own survey that relatively few chemical industry employees had heard of Responsible Care.[55]

The Rotterdam Charter For Sustainable Development

The International Chamber of Commerce finalized its Business Charter for Sustainable Development, also known as the Rotterdam Charter, about a year before the June 1992 Earth Summit. It contains "Principles

for Environmental Management" which are similar in many ways to
Responsible Care, but which add an emphasis on business's role in
creating "sustainable economic development."

The Charter contains revealing clues to the ICC's approach to
economics and the environment, with a convenient but unsubstantiated
assumption that there is a natural convergence between the needs of
environmental protection, sustainable development, economic growth,
and profitable market conditions for business:

> "Economic growth provides the conditions in which protection of the
> environment can best be achieved....In turn, versatile, dynamic, respon-
> sive and profitable businesses are required as the driving force for
> sustainable economic development and for providing managerial,
> technical and financial resources to contribute to the resolution of
> environmental challenges. Market economies, characterized by entre-
> preneurial initiatives, are essential to achieving this."[56]

The ICC ignores both the experience of Southern countries and
analyses which show that the financial resources generated for the
environment by unregulated growth do not catch up with the costs to
people and the environment from so much dirty industrial activity. The
health and environmental crises along the US/Mexican border created by
the explosive growth of the maquiladoras is but one graphic example.

The ICC Charter's definition of the "precautionary approach" is also
revealing:

> "[T]o modify the manufacture, marketing or use of products or services
> or the conduct of activities, consistent with scientific and technical
> understanding, to prevent serious or irreversible environmental degra-
> dation."[57]

The ICC's definition of the precautionary approach is precisely the
opposite of what the principle originally intended, that is, action can and
should be taken to protect the environment even in the absence of
scientific proof. This definition has been supported by international
bodies ranging from UNCED to the Organization for African Unity,

which defines the precautionary approach as "preventing the release into the environment of substances which may cause harm to humans or the environment without waiting for scientific proof regarding such harm."[58]

Conduct Under the Corporate Codes

The ICC may be content to simply modify production, after scientific proof shows that production is causing serious pollution, but the lesson that environmentalists should take from corporate behavior, and from the voluntary codes of conduct, is a sobering one. Even in countries with relatively strong national environmental laws, severe pollution and irreversible ecological degradation are a part of routine business practice, often with catastrophic consequences.

Voluntary codes must be understood in this context. The ICC's "precautionary approach" would fit DuPont's behavior in the case of chlorofluorocarbons nicely; the company waited 14 years after scientists first linked CFCs to ozone destruction before it agreed to stop making them. Only after government and corporate scientists fully "understood" the CFC-ozone layer depletion link in 1992 – long after the public reached an understanding – did DuPont agree to "modify" production, and only then did it replace CFCs with another proven hazardous substance: hydrochlorofluorocarbons (HCFCs). In its publication *From Ideas to Action*, the ICC chose DuPont to illustrate the corporate precautionary approach with the claim that the company's actions "enabled the Montreal Protocol to be ratified."[59] Thus is DuPont's reprehensible role in one of the century's worst environmental catastrophes greenwashed by a perversion of the precautionary approach.

And Beyond the PR?

As a response to people's rising concern about the environmental impacts of TNCs, corporate greenwash is little more than public relations. Yet many people wonder: Don't top corporate managers share the environmental concerns of the rest of the world? Don't at least some of them really want to make more than incremental improvements? The answer is that personal concerns of corporate executives are largely irrelevant.

One reason is that at the individual and corporate level, an ideological emphasis on deregulation and "free" market system nearly always overrides environmental considerations when the two conflict. More immediately, the current market system would often turn an instinct for environmental protection into economic folly for the corporation.

There are a variety of market-based mechanisms which can, in theory, help steer TNCs toward greener products and processes. These include internal budgeting systems, product "eco-labelling," and taxes on natural resource use or toxic emissions. According to Harris Gleckman, however, "there is no agreement on how [internalization of environmental costs] should happen while maintaining the current competitive structures of the market system. Proponents of the free market vociferously assert that the market can take care of the problem without acknowledging that the market has not done so in the past."[60] Meanwhile, they are usually critical of concrete measures to institute market mechanisms for environmental accounting and protection.[61]

For example, some top TNC executives (including those in the WBCSD) have publicly stated their belief in the need for full-cost environmental accounting, whereby prices are made to reflect the costs of natural resources and environmental degradation. But as we have seen, the ICC lobbied to remove environmental accounting measures from UNCED texts.

Moreover, were a firm to decide unilaterally to internalize environmental costs, it would put itself at a disadvantage to its competitors, which continue to externalize such costs. Absent internationally agreed rules, full-cost accounting advocates must first convince even the most hesitant of their competitors to join the cause. International regulation requiring environmental accounting would seem a prerequisite for its implementation, and should logically receive support from corporate environmentalists.

Instead, big business has resisted environmental regulation for its activities in the global arena and has actively promoted the concept of international "self-regulation," including the codes of conduct."Self-regulation," Gleckman correctly points out,"is really an oxymoron. Potential polluters cannot make 'laws' (i.e., regulate) and order 'sanctions' (i.e., authorize penalties and fines) that are against self-interest."[62]

Despite – or because of – this contradiction, the concept of "self-regulation" has become widely accepted by many individual corporations, industry trade associations, and international business trade associations.[63]

If we cannot hope for substantial help from management's personal concerns or corporate self-regulation, the question remains whether or not there are other emerging factors which may affect TNCs bottom line so as to alter their behavior. At present, there is little evidence to support an affirmative answer. In recent years, for example, the idea of "green" taxes on pollution or resource use has appeared in Northern countries. According to the Wuppertal Institute's Dr. Ernst von Weizsacker, a prominent advocate of this taxation: "Green taxes are likely to have an immense steering power. The markets of goods, services and technology would receive an immense stimulus towards environmentally better solutions."[64] Given the fundamental changes they might engender, however, it is unsurprising that the idea of such taxes, notably those aimed at curbing greenhouse gas emissions, has elicited a strong negative response from industry in the United States and Europe.[65] For example, pressure from the manufacturing and oil sectors helped kill the Clinton Administration's proposed broad-based energy tax in 1993.[66]

TNCs cannot ignore fines for environmental damages, but neither is there strong indication that such costs are causing TNCs to change their activities radically. In 1994, for example, a US federal court ordered Exxon to pay US$5 billion for the Valdez oil spill – the largest punitive award ever against a corporation. But Exxon is so big that even if it ends up paying the full amount, industry analysts said the award would not be enough to compel the company to alter its operations, strategies, or dividend policy.[67] "Exxon is the Jupiter of the oil industry," one analyst asserted. "You can have a great comet hit 'em and it barely leaves a scar."[68]

In the US, the issue of informing investors about environmental liability has received some attention. In 1993, the US Securities and Exchange Commission issued new guidelines requiring companies to disclose environmental problems which might lower their shares' value.[69] Yet neither the threat of governmental enforcement actions, nor that of shareholder class-action lawsuits for undisclosed environmental liabilities, appears to have prompted a major change in reporting behavior.

In a 1994 survey of some 200 large publicly-traded corporations, Annual Report, Inc. of Atlanta, Georgia found that a mere nine per cent said that they were devoting significant space to environmental compliance matters in their 1993 annual reports. The remaining corporations indicated that mention of environmental information would be either only "moderately significant" or "not significant"; 36 per cent said there would be no mention of the company's environmental activities whatsoever. "The survey confirmed what we have already observed," commented Annual Report's President, "that the majority of companies are not in a hurry to provide detailed information regarding environmental liabilities or risks."[70] Such is the current situation in the US, where, it should be noted, the reporting requirements for publicly-traded corporations are among the most stringent in the world.

Nonetheless, there remains at least the potential that pressure from within the business or financial community may eventually be brought to bear which helps to mitigate certain harmful TNC activities. Segments of the insurance industry, hit recently by huge property damage claims from violent storms, have begun to consider the possibility that global warming is to blame and may become more actively involved in international climate negotiations to protect their interests.[71] At some point in the future, greater destruction associated with global warming may result in widespread policies (such as the aforementioned energy tax) which negatively impact investors in fossil fuel corporations, and hence compel them to reconsider their investments.[72]

Lender liability for environmental harm by borrowing companies could provide a very powerful incentive for financing of clean production rather than polluting sectors. However, it will take substantial legislative changes to bring this about.[73]

Thus, if currently the answer to the question about reasons why TNCs might be more seriously concerned about their impacts seems to be "no," there is the hope that this could change. In the meantime, however, greenwash continues to dominate both the style and substance of corporate reactions to environment and development crises.

Greenwash Exposed

Indeed, new outlets for greenwash seem to emerge every day. Polluting companies sponsor Earth Day celebrations. Business associations form environmental front groups to do corporate lobbying. Voluntary environmental policies sprout from corporate headquarters. Plastics companies fund recycling curricula for public schools. Thousands of pages touting the wonders of Responsible Care are printed. And advertising executives appear to have no reticence in proceeding from the sublime to the mundane, as they commandeer the poetry of Lao-tzu and Henry David Thoreau, the music of Beethoven and Strauss, the images of marine mammals and panda bears, and the smiles of workers and children in an attempt to improve the image of gasoline, chemicals, pesticides, plastics, smokestacks, or the corporation itself. Greenwash is alive and well and flourishing all over the globe.

It is easy to focus on the obvious absurdity of the ubiquitous environmental advertisements. And it is tragic to contemplate the pernicious effect such ads can have on children, who may grow up believing that oil companies are the main preservers of wilderness or that farmers need synthetic chemicals to work in harmony with nature. But the real political danger of greenwash lies in the possibility it will convince governments and intergovernmental organizations to abdicate further their responsibility to regulate and hold accountable the TNCs.

Even a cursory look at TNC operations clearly demonstrates that TNCs in the chemical, fossil fuel, resource extraction, waste disposal, and nuclear industries have thrived on industrial development strategies which are fundamentally unsustainable. Such corporations, along with multilateral development banks and bilateral aid agencies, are driving forces behind the dominant development model which emphasizes continual growth in production, Gross National Product, free trade, and consumption for the rich.

Now they say they have changed. That they are spending money for the environment. That they will regulate and police themselves. That their technologies are safe. That their projects help the poor. We urge you to look critically at their real-world behavior, starting with the 20 corporations profiled in this book.

Product Stewardship

Product Stewardship is a group of policies adopted by major agrochemical companies which purportedly reduce harm caused by the use of dangerous pesticides. This is the means by which the industry has translated recommendations from the Food and Agriculture Organization's International Code of Conduct on the Distribution and Use of Pesticides into corporate policy. The voluntary FAO Code recommends that pesticides which require special clothing be avoided, that industry should work with governments to solve pesticide problems, and that manufacturers should voluntarily recall products that present unacceptable hazards.

In practice, pesticide companies regularly disregard the Code and continue to market substances that kill and cause long-term damage to people and the environment. Simply put, transnational corporations play a major role in the production of banned and hazardous products, such as pesticides, and therefore have a conflict of interest with the some of the aims of the FAO Code.

German chemical giant Hoechst's behavior in the Philippines provides a stark contrast to the ideals of Product Stewardship which it espouses. Hoechst, the manufacturer of Thiodan (endosulfan), doggedly challenged a ban on endosulfan ordered by the Fertilizer and Pesticide Authority of the Philippines for two years. Hoechst obtained a reversal of the ban from the Regional Trial Court. Not satisfied with this display of its muscle, Hoechst also pursued a civil law suit against

Dr. Romy Quijano, who stated at a workshop about the effects of pesticides on women that Thiodan may cause cancer. The company has also named *Philippines News and Features,* which carried the statement in an article, in the suit. Endosulfan, the leading cause of pesticide poisonings in the Philippines, became the target of a "People's Ban" called by the Pesticide Action Network/Asia and the Pacific.

In 1994, the Philippines Supreme Court upheld the Fertilizer and Pesticide Authority, effectively banning endosulfan except in products with less than five per cent concentration. Even then, Hoechst attempted to intervene, with two letters to the Philippine President requesting the formation of a new arbitration committee and saying that their "decision to remain in the Philippines [is] at stake."

The Hoechst case demonstrates that large TNCs, singly and collectively, sometimes possess greater resources and wield more power than national regulatory authorities. International bans on the most hazardous products would restore a balance of power between corporations, governments, and private citizens by limiting TNCs' ability to influence and overwhelm governmental and non-governmental efforts to eliminate hazardous chemicals at the national level.

(**Sources**: Barbara Dinham, *The Pesticide Hazard: A Global Health and Environmental Audit,* Zed Books, London 1993; *CITIZENS PESTICIDES HOECHST The Story of Endosulfan and Triphenyltin,* Ronald Macfarlane, Pesticide Action Network Asia and the Pacific, Malaysia 1994; "International Citizens' Campaign Targets Hoechst Pesticides," *Global Pesticides Campaigner,* September 1994.)

Endnotes

1. Remarks before the National Wildlife Federation's Synergy Conference, January 1990, quoted in Jack Doyle, Friends of the Earth (FOE), *Hold the Applause! A Case Study of Corporate Environmentalism,* Washington, DC, August 1991.
2. Vandana Shiva, "Transfer of Technology," Third World Network Briefing Papers for UNCED, August 1991.
3. Doyle, op cit 1, p. 57.
4. Meera Nanda, "Waiting for Justice—Union Carbide's Legacy in Bhopal," in *Multinational Monitor,* July/August 1991, p. 16. See also Sanjay Kumar, "Union Carbide officials face prosecution," *New Scientist,* 1 May 1993.
5. Molly Moore, "In Bhopal, a Relentless Cloud of Despair," *The Washington Post National Weekly Edition,* 4-10 October 1993.
6. Josh Karliner, "The Bhopal Tragedy: Ten Years After," *Global Pesticide Campaigner,* December 1994.
7. Lis Wiehl, "Texas Courts Opened to Foreign Damage Cases," *The New York Times,* 25 May 1990.
8. Business Council for Sustainable Development, background information, 1991.
9. Herman Daly & Robert Goodland, the World Bank Environment Department, "An Ecological-Economic Assessment of Deregulation of International Commerce Under GATT," Spring 1993 Draft, p. 14.
10. *Environmental Aspects of the Activities of Transnational Corporations: A Survey,* United Nations Centre on Transnational Corporations, United Nations, New York, 1985, p. 93.
11. Nanda, op cit 4, p. 15, and Karliner, op cit 6.
12. "Die Katastrophe als Gluckfall," in *Weltwoche,* 31 October 1991. Also Greenpeace private communication with Resende plant employee.
13. Dr. Arjun Makhijani, and Dr. A. van Buren, A. Bickel, S. Saleska, *Climate Change and Transnational Corporations Analysis and Trends,* United Nations Centre on Transnational Corporations, Environment Series No. 2, United Nations, New York, 1992, p. 47.
14. "Ongoing and Future Research: Transnational Corporations and Issues Relating to the Environment," United Nations Commission on Transnational Corporations (UNCTC), 5-14 April 1989, p. 6.
15. Makhijani et al, op cit 13, p. 77.
16. Steven Shrybman, "The Environmental Costs of Free Trade," in the *Multinational Monitor,* March 1990, p. 20. See also "Transnational Corpora-

tions and Issues Relating to the Environment: The Contribution of the Commission and UNCTC to the Work of the Preparatory Committee for the United Nations Conference on Environment and Development," UNCTC, 28 February 1991; and "Activities of the Transnational Corporations and Management Division and Its Joint Units," E/C.10/1993/7, Commission on Transnational Corporations, 19th session, 5-15 April 1993, p. 12.

17. Joe Thornton, *The Product is the Poison – A Case for Chlorine Phase-Out,* Greenpeace USA, Washington, DC, 1991.

18. From "Pesticides: Export of Unregistered Pesticides is not Adequately Monitored by the EPA," General Accounting Office of the US Congress, April 1989. Cited in "Unregistered Pesticides Rejected Toxics Escape Export Controls," Greenpeace International and Pesticide Action Network-FRG, October 1990, p. 2.

19. Personal communication of Kenny Bruno with Ethyl representative.

20. For the information about TNCs and FDI see: "Trends on Foreign Direct Investment – Report by the UNCED Secretariat," Commission on Transnational Corporations, 20th session, E/C.10/1994/2, 11 March 1994; *World Investment Report 1994 – Transnational Corporations, Employment and the Workplace*, UNCTAD Division on Transnational Corporations and Investment, Geneva, 1994, chapter 1; "Ongoing and Future Research: Transnational Corporations and Issues Relating to the Environment," United Nations Centre on Transnational Corporations, 5-14 April 1989, p. 5; "Activities of the Transnational Corporations and Management Division and Its Joint Units," E/C.10/1993/7, Commission on Transnational Corporations, 19th session, 5-15 April 1993, p. 12; and "Trends in Foreign Direct Investment and Transnational Corporations, 21st session, 24 April 1995, TD/B/ITNC/2.

21. See "Trends in Foreign Direct Investment," General Discussion on Transnational Corporations in the World Economy and Trends in Foreign Direct Investment in Developing Countries," E/C.10/1993/2, Commission on Transnational Corporations , 19th session, 5-15 April 1993, p. 8; *World Investment Report 1992 Transnational Corporations as Engines of Growth,* Transnational Corporations and Management Division, United Nations, New York, 1992, pp. 1-2; and *Transnational Corporations in World Development Trends and Prospects,* United Nations, New York, 1988, pp. 3-4.

22. *World Development Report 1994*, op cit 20.

23. Jeffrey Sachs, "Beyond Bretton Woods – A new blueprint," *The Economist,* 1-7 October 1994, p. 24.

24. Richard J. Barnet & John Cavanagh, *Global Dreams: Imperial Corpora-*

tions and The New World Order, New York, Simon & Schuster, 1994, p. 341.

25. Ibid, p. 348.

26. Nicholas Hildyard, "Maastricht: The Protectionism of Free Trade," *The Ecologist*, v. 23, March/April 1993; Sarah Anderson, John Cavanagh, & Sandra Gross, *NAFTA's Corporate Cadre An Analysis of the USA*NAFTA State Captains,* The Institute for Policy Studies, Washington, DC, July 1993; and Chakravarthi Raghavan, *Recolonization – GATT, The Uruguay Round & The Third World,* London, Zed Books, and Penang, Third World Network, 1990, esp. pp. 62-63, 130-155.

27. Daly & Goodland, op cit 9, p. 45.

28. Richard Stevenson, "ABB Asea Brown Boveri – Global Arms, Regional Muscle," *The New York Times*, 3 January 1994.

29. "EPA Head Defends NAFTA as Greenest Trade Pact Ever," Reuters, 13 August 1992.

30. See, for example, Tim Lang and Colin Hines, *The New Protectionism – Protecting the Future Against Free Trade,* London, Earthscan, 1993, passim.

31. Noam Chomsky, "Notes on NAFTA," *The Nation*, 29 March 1993.

32. Ibid.

33. Barnet & Cavanagh, op cit 24, p. 422.

34. Ibid, p. 341.

35. Harris Gleckman & Riva Krut, Benchmark Environmental Consulting, "Business Competition and Competition Policy – The Case for International Action," Christian Aid, London, June 1994, p. 4.

36. The Heritage Foundation, "A United Nations Assessment Project Study," 24 May 1991, p. 2. Also personal communication of Kenny Bruno with Harris Gleckman.

37. "TNCs: Sweden, Norway Refuse to Fight, in "Crosscurrents," an independent NGO newspaper for UNCED, 30 March - 1 April, 1992.

38. Personal communication of Kenny Bruno with Kristin Dawkins. For an example, see the "World Summit for Social Development Declaration and Programme for Action," United Nations, 1995.

39. For a fuller exploration of technology transfer issues in UNCED, see "Technological Transformation," submission by Greenpeace International to the Fourth Preparatory Negotiation for UNCED, March 1992.

40. "Agenda 21," Chapter 34.1, "Transfer of Environmentally Sound Technology, Cooperation, and Capacity-Building," p. 2.

41. Kenny Bruno, "The Corporate Capture of the Earth Summit," *Multinational Monitor*, July/August 1992.
42. Ibid.
43. Kristin Dawkins, "NAFTA, GATT & The World Trade Organization — The Emerging New World Order," Open Pamphlet Magazine Series, Westfield, New Jersey, 1994, pp. 7-8.
44. See Klaus Vieli, "Stephan Schmidheiny," *Das Magazin*, 16 March 1990, and Andre Carrothers, "The Merchants of UNCED," *Multinational Monitor,* July/August 1992, pp. 18-19.
45. BSCD press release "Two Leading Business Organizations to Merge," 17 November 1994.
46. Milton Moskowitz, Michael Kaz, and Robert Levering, eds., *Everybody's Business: An Almanac,* New York, Harper & Row, 1980, and Joyce Nelson, "The great global greenwash – Burson-Marsteller vs. the Environment," *Third World Resurgence,* no. 37, September 1993, p. 8. This article was originally published in *Covert Action Quarterly*, no. 44, Spring 1993.
47. For this and the information about A.H. Robbins and Union Carbide, see Nelson, ibid.
48. Nelson, op cit 46. See also Joyce Nelson, *Sultans of Sleaze: Public Relations and the Media*, Toronto, Between The Lines, 1989, pp. 22-42.
49. From Burson-Marsteller brochure quoted in Nelson, op cit 46, p. 8.
50. *Journal of Commerce,* 23 December 1988; *La Liberte,* (Switzerland), 14 November 1988.
51. Barnet & Cavanagh, op cit 24, p. 341.
52. David Hunter, "Staying Off the Scrap Heap and Preparing for the Next Century," *Chemicalweek,* 12 May 1993.
53. For information on Responsible Care, see "Responsible Care A Public Commitment" and "Responsible Care Progress Report 1991," both published by the Chemical Manufacturers Association, Washington, DC. See also *Chemicalweek's* Special Issue on Responsible Care, 17 July 1991, especially p. 9-32.
54. Emily S. Plishner, "Environmental Activists Question Responsible Care," *Chemicalweek,* 1 April 1992.
55. Karen Heller, "Public Opinion: Still Not High," *Chemicalweek,* 23 September 1992.
56. See the Business Charter for Sustainable Development, adopted by the 64th session of the International Chamber of Commerce Executive Board on 27 November 1990 and first published in April 1991. Prepared by the ICC

Commission on Environment Working Party on Sustainable Development, Paris.

57. Ibid.

58. Bamako Convention, Article 4, paragraph 3(f).

59. International Environmental Bureau of the ICC, *From Ideas to Action – Business and Sustainable Development, The ICC Report on the Greening of Enterprise 92*, Special Edition for the UN Conference on Environment and Development, Oslo, Norway, 1992, p. 199.

60. Harris Gleckman, "Transnational Corporations: Strategic Responses to 'Sustainable Development'," 12 May 1994 draft, p. 18.

61. Ibid.

62. Ibid, p. 12.

63. Gleckman & Krut, op cit 35, p. 3.

64. Ernst von Weizsacker, "Global Challenges and Environmental Tax Reform," International Conference on Economic Instruments for Environmental Protection, Rome, 20 January 1990, p. 7.

65. See Bob Woodward, *The Agenda– Inside the Clinton White House*, Simon & Schuster, New York, 1993, esp. pp. 217-221, and Ann Doherty & Olivier Hoedeman, "Misshaping Europe: The European Roundtable of Industrialists," *The Ecologist*, v. 24, no. 4, July/August 1994.

66. Woodward, ibid, p. 217.

67. Caleb Solomon, "Exxon Is Told to Pay $5 Billion For Valdez Spill," *The Wall Street Journal*, 19 September 1994.

68. Ibid.

69. Annual Reports, Inc. 22 February 1994 press release "Survey Shows Most Annual Reports Will Provide Very Little Information on Environmental Compliance."

70. Ibid.

71. Eugene Linden, "Burned by Warming," *Time*, 14 March 1994.

72. For a discussion of this issue see Mark Mansely, The Delphi Group, "Long Term Financial Risks to the Carbon Fuel Industry from Climate Change," Delphi International Ltd., London, 1994.

73. Kaspar Muller et al, "Environmental Reporting and Disclosures – The Financial Analyst's View," published by the European Federation of Financial Analysts' Societies, September 1994, p. 2.

Source: Cartoon by Kirk Anderson.

Protected
by Shell

Caring for the world is a responsibility we all share. And it's one we take seriously at Shell. We started the Better Environment Awards Scheme, and we support the World Environment Day, the Tree Project, and other efforts that will benefit generations to come.

In fact, as long as the earth needs someone to care for it, you can be sure of Shell.

Source: *One Earth*, Friends of the Earth (Charity) Ltd., Hong Kong, Autumn 1991.

Greenwash Snapshot #1

ROYAL DUTCH/SHELL GROUP

A case study in global warming, oil pollution, and pesticide poisoning.

Royal Dutch/Shell Group (Shell)
Group Chairman: Sir Peter Holmes
Headquarters: 30 Carel van Bylandtlaan, 2596 HR The Hague, NV
Telephone: 31-70-377-4540 Fax: 31-20-377-4848
Shell Headquarters: Shell Centre, London SE1 7NA, United Kingdom
Telephone: 44-171-934-3856 Fax: 44-171-934-8060
Major businesses: oil; natural gas; chemicals; coal; and metals.
Major subsidiary: Shell Oil Company (Houston, Texas US).

Royal Dutch/Shell has signed the ICC Rotterdam Charter and is a member of the WBCSD.

The World's Largest Non-government Oil and Natural Gas Company

Royal Dutch/Shell explores for oil in some 50 countries, refines in 34, and markets in over 100 nations. The company owns 400 million acres of land in 50 countries, employs 133,000 people, has annual sales of around US$100 billion, and US$9 billion in cash reserves.[1] Shell's operations include eucalyptus plantations in Asia, Latin America, and Africa, and bauxite projects in Brazil.

With gross revenues of over US$102 billion in 1991 – bigger than the gross domestic products of most countries – Shell could be a powerful force for a transition to sustainable energy systems and economies all over the globe. Instead, this mammoth concentration of resources is dedicated mainly to the world's largest unsustainable industry: OIL.

Shell, The Precautionary Approach, and Global Warming

The burning of oil is responsible for about 40 per cent of the energy-related carbon dioxide added to the atmosphere annually, and oil production and consumption contribute to about 26 per cent of greenhouse gases. Shell activities, which include handling about eight per cent of the world's oil and natural gas, account for approximately three per cent of the human contribution to carbon dioxide emissions, without counting their interests in coal and other greenhouse gas sources.[2]

The possible consequences of global climate change include changes in rainfall patterns, bleaching of coral reef systems, mass extinction of species, disruption of food sources, rising sea levels, and flooding of coastal plains and islands which could force millions of people from their homes. As diplomat Robert van Lierop of the Pacific island nation of Vanuatu put it: "It's a question of survival, it's that simple. At the very least, sea level rises of a foot or so would wipe out island ecosystems. At worst, whole islands could disappear under water."[3]

The magnitude of the threat from continued dependence on oil is monumental in the best-case scenario, unimaginable in a worst-case scenario. In theory, Shell recognizes the threat from global warming and the need for changes in the oil business: "Shell companies...believe that there is enough indication of potential risk to the environment [from global climate change]...to start to adopt precautionary measures."[4]

Yet despite this ostensible commitment to a precautionary approach, Shell plays an aggressive role in the drive to develop the world's one trillion barrels of known oil reserves and the search for the world's unlocated 750 million or so barrels, investing nearly US$842 million in research and development during 1990.

The company will not disclose how much it allocates for biomass or solar research, as opposed to the traditional businesses of oil, natural gas, and coal. But Shell does say that it spent US$5 billion – more than three-quarters of its net earnings that year – on oil and gas exploration and production in 1990 alone, with new ventures starting in Algeria, Guatemala, Kenya, the Philippines, and Yemen.[5] Shell justifies this type of investment with: "World energy needs will likely require a better than 50 per cent increase in the fossil fuel production rate over the next 40 years..."[6]

In the context of its role in oil development and global warming, Shell's emphasis on "environmental performance" and its embrace of "precautionary measures" appear to be pure greenwash. The company also continues to support an industry lobby group in the United States, the Global Climate Coalition, which has stated that "existing scientific evidence does not support actions aimed solely at reducing or stabilizing greenhouse gas emissions."[7]

Apparently, however, Shell feels that global warming is a serious issue where its own investment is at stake. In 1990, the company's Norwegian unit A/S Norske reported it would add a meter to the height of the North Sea Troll gas platform to take into account projected sea level rises from global warming.[8]

Oil Production Means Oil Pollution

At every stage of the oil life-cycle – exploration, production, transportation, manufacturing, and consumption – there can be enormous damage to the environment and human health. A brief look at some of Shell's recent experiences is telling:

- In May 1988, seven Shell employees were killed at an explosion at the company's Norco, Louisiana refinery. Shell has paid at least US$24 million in damage claims.[9]

- In April 1988, a spill of 440,000 gallons of crude oil at Shell's Martinez, California refinery polluted over 100 acres of wetlands and 11 miles of shoreline, killing hundreds of animals and costing the company US$20 million in penalties and US$12 million in cleanup bills.[10]

- In 1989, a Shell refinery in the United Kingdom spilled 10,000 gallons of crude oil into the Mersey River. Shell was fined US$1.6 million and paid another US$2.24 million for cleanup costs.[11]

- In the fall of 1989, a Shell tanker spilled enough oil near the island of St. Lucia in the Caribbean to cover Bannes Bay for two weeks. The company refused to indicate publicly how it planned to prevent such accidents in the future, but it did participate in an anti-litter campaign as part of St. Lucia's Heritage Week.[12]

- In 1991, The US Environmental Protection Agency (EPA) cited Shell and nine other major oil companies for discharging contaminated fluids from service stations into or directly above underground sources of water. The company paid US$56,000 and said it would clean up the sites by 1993.[13]

Shell in Nigeria

"As the developing world strives to meet the growing needs of its people, to industrialize and develop an industrial base and create market economies, there will be significant demand for energy....This growth in energy demand is almost a prerequisite to achieving sustainable development. If we accept that growth is required to meet the natural aspirations of the people of developing countries and to provide the inputs to achieve sustainable development, we need to recognize that achieving such growth will require the consumption of increasing quantities of energy, primarily fossil fuel energy, and that this will need to be made available by companies such as ourselves."
  – *Royal Dutch/Shell Group spokesperson, 1993*[14]

Oil companies have operated in southern Nigeria on lands of people such as the Ogoni since 1958. Shell holds more reserves than any other oil company in Nigeria, and the country accounts for 14 per cent of Shell's global oil production. The Nigerian government receives 80 per cent of its revenues from oil, with Shell's production accounting for half those revenues.[15] But the experience of Shell in Nigeria does little to support the company's contention that in less-industrialized countries' increased energy demand supplied by the fossil fuel industry leads naturally to sustainable development, or fulfils people's "natural aspirations."

The oil TNCs' activities have caused extensive pollution in the Niger Delta. Oil pipelines run above ground through villages and agricultural areas, wells are located within village boundaries, and gas is flared continuously near human settlements.[16] As a result, the Ogoni, whose poverty is dire even by African standards, are losing their ability to farm and fish locally and many Ogoni youth are leaving their home to seek a livelihood elsewhere. According to Ken Saro-Wiwa, a prominent Nigerian activist who was imprisoned and later executed by the Nigerian government for what many believe was his leadership role in confronting Shell, if the present practices of the oil companies continue, the Ogoni community will be gone within three decades.[17]

The Ogoni have long challenged the oil companies, especially Shell, over the ecological degradation and its social consequences. The Ogoni have demanded from the oil producers payment of US$10 billion in royalties and compensation for the environmental harm, and also want the TNCs bury their pipelines and end gas flaring. It is estimated that Ogoni land has contributed some US$30 billion in oil revenues.[18]

The conflict between the people of the Niger Delta and the oil companies has led to violence. In 1990, the police massacred 80 unarmed villagers – including an Ogoni leader – after Shell asked for help against demonstrators demanding compensation for lost lands. Shell said it regretted the massacre.[19] A subsequent report into the killings by a Nigerian judicial commission, which criticized Shell's activities and accused the police of atrocities, was suppressed and had to be smuggled out of the country. The report recommended that much more revenue from oil should be given to people whose land had been taken from them. It also said Shell should join a tribunal working locally to settle disputes and pay compensation.[20]

In January 1993, over 100,000 Ogoni people demonstrated against the activities of Shell and other oil producers, focusing upon the companies' continued unwillingness to address Ogoni demands.[21] Since then, the situation has worsened, with Nigerian soldiers attacking and destroying Ogoni villages. The Ogoni say the raids are retribution by the government for confronting Shell.[22]

"I curse the day Shell found oil on our land."
—   *Grace Zorbidom, an Ogoni, 1994* [23]

The DBCP Case

Shell Oil was a co-defendant in one of the most important hazardous chemical export liability cases ever. For over two decades, Shell supplied a pesticide containing dibromochloropropane (DBCP) to Standard Fruit Company for use in banana plantations. Shell had known since the 1950s that DBCP causes sterility in male laboratory animals, but did not include this information on product labels. Even after the US EPA determined that DBCP caused sterility in humans and banned production, Shell continued to market the chemical. [24]

Allegedly after suffering exposure to DBCP, between 500 and 2,000 banana workers in Costa Rica became sterile and today continue to face higher cancer risk. Depression, alcoholism, suicide, and divorces have all increased among the banana workers since their exposure to DBCP.[25]

In 1984, Costa Rican workers filed suit against Shell and Dow Chemical, also a manufacturer of DBCP (a second suit was filed in 1991). In response, the companies pleaded "forum non conveniens," arguing that it was too inconvenient to hear the Costa Ricans' case in the United States. The Texas Supreme Court denied this motion in 1990 in what was hailed as a landmark decision. On Shell and Dow's legal manoeuvres, the Court's Justice Lloyd Doggett wrote: "What is really involved is not convenience but connivance to avoid corporate accountability."[26]

In 1992, 981 of the Costa Rican workers reached a settlement agreement with Shell and Dow for a reported US$20 million.[27] In 1993, another suit in Texas against Shell and Dow about DBCP exposure began on behalf of over 16,000 workers, mostly from Central America but also from the Caribbean, Africa, Asia, and South America.[28]

Overall, Shell is one of the world's largest pesticide manufacturers, the inventor and a producer of dieldrin and aldrin – two of the so-called "Dirty Dozen" hazardous pesticides. Both "drins" have been banned in over 35 countries.[29] Shell has now eliminated production of the "drins."

"Anyway, from what I hear they could use a little birth control down
there."
– *Clyde MacBeth, former Shell scientist, speaking
about the Costa Rican DBCP sterility issue, 1989*[30]

Endnotes

1. *Hoover's Handbook of World Business 1993*. The Reference Press, Inc.,
Austin, Texas, 1993. p. 410.
2. For more information see Jeremy Leggett, "Environmental Responsibilities
in the Oil and Gas Industry: A View From the Environmental Movement,"
Paper for the First International Conference on Health, Safety, and Environment in Oil and Gas Exploration and Production, Netherlands Congress
Centre, The Hague, November 1991. According to this source, 60 per cent
of the human-enhanced greenhouse effect comes from the production and
consumption of fossil fuels, mostly as carbon dioxide derived directly from
combustion (p. 4). Of that, 44 per cent (1989 figure), or 26 per cent of the
total, comes from oil (p. 21, note 10). Shell claims it contributes about 0.5
per cent to global emissions of CO_2, but does not explain how it arrived at
this figure.
3. Paul Lewis, "Island Nations Fear a Rise in the Sea," *The New York Times*,
17 February 1992. For more on the effects of climate change see Greenpeace
International, "Climate Time Bomb – Signs of Climate Change from the
Greenpeace Database," Amsterdam, 1994.
4. "The Shell Review," June 1991. pp. 53-54.
5. "The Shell Review," p. 18.
6. Quoted in Leggett, "Environmental Responsibilities in the Oil and Gas
Industry," p. 8.
7. GCC Press Release, 19 February 1992.
8. Reuters, 1 February 1992.
9. From The Council on Economic Priorities (CEP) Corporate Environmental
Data Clearinghouse, "Environmental Profile of the Royal Dutch Petroleum
Company, The Shell Transport and Trading Company, p.l.c., Shell Oil
Company," 1991. p. 24.
10. Ibid, pp. 18-19.
11. Ibid, p. 30.
12. Earl Bousquet, "St. Lucia: Oil Transnationals Working on Environmental
Image," Inter Press Service International News, 14 November 1989.

13. Philip Mattera, *World Class Business – A Guide to the 100 Most Powerful Corporations*, Henry Holt and Company, New York, 1992, p. 582.
14. Quoted in *New World Journal*, May 1993.
15. Geraldine Brooks, "Shell's Nigerian Fields Produce Few Benefits for Region's Villagers," in *The Wall Street Journal Europe*, 6-7 May 1994.
16. Paul Brown, "80 Nigerians killed in Shell oil protest – Cover-Up charge as villagers demand compensation," *The Guardian*, 8 October 1992.
17. Quoted in Greenpeace press release, London, 29 July 1993.
18. "Oil Protest in Nigeria," in "Campaigns" section, *The Ecologist*, vol. 23, no. 2, March/April 1993.
19. Brown, op cit.
20. Brown, op cit.
21. "Oil Protest in Nigeria," op cit.
22. Brooks, op cit.
23. Brooks, op cit.
24. "DBCP: The Legacy," produced by Misko, Howie, & Sweeney, Dallas, Texas, 1994, pp. 10-11.
25. David Weir and Constance Matthiessen, "Will the Circle Be Unbroken?," in *Mother Jones*, June 1989, p. 21-27.
26. Lis Wiehl, "Texas Courts Opened to Foreign Damage Cases," *The New York Times*, 25 May 1990. See also Patricia Crisafulli, "Costa Ricans File Suit Faulting Pesticides," in *Journal of Commerce*, 22 July 1991.
27. "Costa Rican Banana Workers Suing Standard Fruit, Dow Chemical, and Shell Oil Settle Out of Court for More than $10 million," *Environment Watch Latin America*, September 1992; and "Costa Rica Workers' Lawsuit Alleging that DBCP Caused Sterility in Close to Settlement," in "Corporate Crime Reporter," 20 July 1992.
28. Personal communication with representative of law firm Misko, Howie, & Sweeney, May 1993. Also see their "DBCP: The Legacy."
29. See Global Pesticide Campaigner, vol. 1, no. 3, June 1991. p.7. Updated from the United Nations' *Consolidated List of Products whose Consumption and/or Sale Have Been Banned, Withdrawn, Severely Restricted or Not Approved by Governments*, 5th edition, 1994.
30. Quoted in Weir & Matthiessen, op cit, p. 24.

' AT the root of my travails lies Shell, which has exploited, traduced, and driven the Ogoni to extinction in the last three decades. The company has... left a completely devastated environment and a trail of human misery. When I organized the Ogoni people to protest peacefully against Shell's ecological war, the company invited the Nigerian military to intervene... I have one suggestion for those whose conscience has been disturbed by my story: boycott all Shell products. Picket Shell garages. Do not allow them to profit by their destruction of the people and ecology of the Niger delta... Support the call for a worldwide boycott of Nigerian oil. Help save ' life and the environment of the Niger delta.

Source: Quote from Ken Saro-Wiwa, reprinted in Andy Rowell, "Oil, Shell and Nigeria," *The Ecologist*, November/December 1995.

Source: *Mobil World*, Special Issue, March 1991.

Greenwash Snapshot #2

MOBIL CORPORATION

A case study in oil pollution, a biodegradability scam, "green collar fraud," and sham recycling.

Mobil Corporation (Mobil)
Chief Executive Officer: Lucio Noto
Headquarters: 3225 Gallows Road, Fairfax, Virginia 22037, USA
Telephone: 703-846-3000 Fax: 703-846-4669
Major businesses: oil and gas; chemicals; plastics.

Mobil, world's fourth largest petroleum company, explores for and produces oil in 20 countries, and sells petroleum products in over 90. Sixty one per cent of Mobil's production is outside the US; the company derives 66 per cent of its revenue from outside the US as well. Mobil is a signer of Responsible Care and the ICC Rotterdam Charter. It is a member of the WBCSD.

[M]arketing is the part of Mobil's business that's most visible to the public. Most people...see advertising. This presents Mobil's marketers with a unique opportunity to deliver their environmental message – an opportunity they've seized."
      – *From Mobil World, 1991*[1]

Mobil devotes huge amounts of money to market an "environmental message"; its ads in the editorial sections of major US newspapers alone cost hundreds of thousands of dollars annually. These ads typically downplay the destructive impact of Mobil's activities, advocate oil exploration in pristine ecosystems such as the Arctic National Wildlife Refuge, bemoan legislative limits to fossil fuel development and dependency, and extol the environmental virtues of plastics and styrofoam. In an extreme example of "image advertising," the ads sometimes stray into areas far from Mobil's business expertise, like a wandering idyll on American summer memories that appeared in *The New York Times* in August 1992.

Elsewhere, slick company pamphlets tell of Mobil's protection of wildlife habitats, public education on issues like plastics recycling, and environmental grants and awards. Even irrelevant human interest stories, like the one about the Mobil employee in Japan who helped revive his local Little League baseball team by collecting aluminum cans, are not safe from Mobil's green propaganda team. Mobil greenwash is so pervasive and extends to so many issues that we can only scratch the surface in this snapshot.

"Green Collar Fraud": Mobil and the Biodegradability Scam

In 1988, Mobil began advertising its Hefty brand trash bags, made from the plastic resin polyethylene, as biodegradable (that is, capable of being fully metabolized by microorganisms and assimilated into the natural biological cycle in soil and water).[2] This claim, which Mobil has never been able to verify scientifically, was an attempt to cash in on public environmental concern, a fact which a company spokesperson acknowledged in 1989:

> "[Degradable bags] are not an answer to landfill crowding or littering....
> Degradability is just a marketing tool.... We're talking out of both sides
> of our mouths because we want to sell bags. I don't think the average
> consumer even knows what degradability means. Customers don't care
> if it solves the solid-waste problem. It makes them feel good."[3]

Seven states filed lawsuits against Mobil in 1990 for this blatantly cynical marketing scam. Calling Mobil's scam "green collar fraud," New York State Attorney General Robert Abrams summed up the prevailing sentiment:

"It is a myth that degradable plastics quickly disappear or provide any benefits to the environment in a landfill, and Mobil knew it. But that didn't stop Mobil from trying to mislead consumers into purchasing products that simply are not good for the environment....Mobil's claims for its trash bags should be thrown into the landfill of rhetoric."[4]

By 1991, Mobil had settled with the states, paying a total of US$165,000 and agreeing to remove the word degradable from its Hefty bags. In July 1992, Mobil entered into a consent agreement with the US Federal Trade Commission (FTC) which prohibited the company from making further unsubstantiated claims about the Hefty bags. One day after Mobil's settlement, the FTC announced national guidelines for defining terms recyclable and biodegradable. Although voluntary, they are the most specific guidelines the US government has issued on what constitutes misleading environmental advertising.[5]

However, Mobil's biodegradability scam did not hurt sales. In fact, the company's scam coincided with a dramatic rise in the market price of its chemical products such as polyethylene. This helped boost Mobil's gross revenues by over US$1 billion, and net income by US$600 million, from 1987 to 1988. Between 1988 and 1990, the company's earnings from this business segment were at a record high. According to Mobil, "Hefty remains one of the top brands in the consumer marketplace."[6] In late 1995, Tenneco announced it would acquire Mobil's plastic division, including the Hefty brand name trash bags.

Environmental Effects of Oil

Mobil points proudly to its environmental grants program. Among the most significant, says the company, was the grant for a "landmark study" on global warming. Mobil ignores the oil industry's huge contribution to the problem of global warming and the unsustainability of continued oil exploration and development (for more on the relationship between oil and global warming see Shell case study).

Gulf of Mexico

Mobil's 1991 Annual Report illustrates "environmental excellence" with an underwater photograph of fish swimming in blue waters of the Gulf of Mexico, a Mobil "production stronghold." Lurking behind the fish are the legs of an offshore oil rig. Rigs, the company says, "attract and shelter marine life." [7] Left out is any mention the threat oil drilling poses to aquatic species, some of which are susceptible to harm from toxic petroleum products at levels as low as one to 100 parts per billion.[8]

In the Gulf, Mobil and other oil companies discharge daily 1.5 million barrels of "toxic brine" tainted with chemicals and heavy metals that can concentrate in tissue of marine organisms. Their drilling has generated millions of tons of muds and cuttings that can smother bottom-dwelling life. This degradation, plus that from rigs' air pollution, tanker traffic, and spills, affects not only ocean but coastal ecosystems. In Louisiana's coastal plain and barrier islands, for example, wetland loss is occurring at a rate of 50 square miles per year.[9]

Nigeria

For over two decades, seven transnationals have monopolized oil development in Nigeria. Mobil's reserves, which are second only to Shell's, have produced more than 1.5 billion barrels of oil since 1970. Mobil expects to double its current production volume during the next four years with new oil fields off Nigeria's coast. The result of the oil industry's operations, according to Evans Aina, Director of Nigeria's Federal Environmental Protection Agency, has been extensive degradation of the country's land and marine ecosystems. Most of the oil pollution, Aina says, has come from "improper disposal of drilling muds, shipping and terrestrial traffic accidents, tank washing and oil ballast discharges, depot leakage and failure or rupture in oil pipelines." In all, the oil industry in Nigeria has been responsible for at least 2,796 spills or releases of some 2.1 million barrels of oil between 1976 and 1990 (see Shell case study for more on oil TNCs in Nigeria).[10]

California and New York

Like all major oil companies, Mobil has a significant pollution record. The US Environmental Protection Agency (EPA) has named the company a potentially responsible party at 179 Superfund sites. Since the early 1980s, Mobil has spent US$13 million in Superfund or equivalent state legislation expenses. [11] Nowhere is Mobil's record more dismal than in California, location of the company's largest US oil reserves.

In 1988, Mobil's pipelines in Los Angeles ruptured, spilling 130,000 gallons of oil that contaminated the Los Angeles River and killed hundreds of fish and dozens of birds. According to the LA Department of Transportation this was the sixth such rupture since 1973. After the city charged Mobil with negligent maintenance, the company finally replaced 75 miles of leaking California pipeline in 1990. [12] Mobil's refinery in Torrance, near Los Angeles, has experienced many accidents, spills, and violations. From 1987-1989, four major explosions and several fires at the facility killed two workers and injured 15. The Torrance City Council ordered an independent review of the refinery which concluded that extreme carelessness and failure to follow safety guidelines were to blame. Between 1988-1989, Mobil paid US$34,000 in Occupational Safety and Health Administration fines for 105 safety violations. [13]

Concern over the facility's operation prompted the city of Torrance to sue Mobil in 1989 to establish municipal regulatory authority over the refinery. In particular, city leaders feared an uncontrolled release of highly toxic hydrofluoric acid might cause a "disaster of Bhopal-like proportions." [14] Mobil settled the suit by agreeing to stop using hydrofluoric acid if it could not adequately control a release and by hiring a safety consultant.

Mobil paid an US$85,000 penalty and US$3 million in charges in August 1989 to clean up 2.4 million gallons of gasoline which had leaked from tanks and pipelines under Torrance. The gas contained high levels of benzene and other carcinogenic chemicals. During the fall of 1989, the South Coast Air Quality Management District cited the Torrance refinery as having the region's worst record in air quality violations and complaints and banned Mobil from monitoring its own air emissions – a

privilege extended to all other major oil companies in Southern California.[15]

In 1990, the California Department of Health (DOH) fined Mobil
US$66,000 for hazardous waste violations at the Torrance refinery. The
previous year the DOH had fined Mobil US$125,000, and ordered a
US$225,000 cleanup, for similar problems at the facility.[16]

> "Environmental conservation **continues** to be an important objective
> in California."
> –   *From 1991 Mobil Fact Book*[17] (emphasis added)

Mobil's Greenpoint terminal in New York City rivals Torrance for
pollution. Although it does not accept complete responsibility, Mobil
agreed in 1990 with the NY State Department of Environmental Conservation (NYSDEC) to clean up what is believed to be the largest oil leak
in US history underneath the terminal. For the last 40 years, up to 17
million gallons of oil have been released or spilled over some 52 acres,
causing sewer system shutdowns, construction problems, and threats to
underlying groundwater. State officials estimate that the cleanup will cost
Mobil tens of millions of dollars.[18]

Mobil has created other hazards at Greenpoint. In 1988, after a 60,000
gallon leak of gasoline at the terminal, Mobil spent US$600,000 to move
pipelines above ground and to build new storage tanks and monitoring
wells. In 1990, 50,000 gallons of kerosene leaked from the terminal.
Because Mobil failed to report the release, NYSDEC fined the company
US$500,000, the maximum penalty possible.[19]

Mobil agreed in 1988 to purchase eight homes in Jacksonville, NY
that had been contaminated by leaking underground gasoline tanks. The
homeowners had been exposed to carcinogenic benzene. In 1990, a Mobil
pipeline in West Seneca, NY burst and spilled 20,000 gallons of gasoline.
Over 300 families had to be temporarily evacuated.[20]

Economic Effects

Besides ecological harm, oil development can also create economic
havoc. Production in the Gulf of Mexico has injured the tourist trade

while commercial fisherfolk have lost fish as well as fishing grounds. Moreover, boom-and-bust cycles typify oil operations in the Gulf as elsewhere, making stable employment impossible. "Sure," a Gulf worker said after a boom period, "there was lots of money to be made at first, but as soon as the price of oil dropped, the flow of cash dried up and the oil companies moved on.... Suddenly, there were no jobs, just pollution."[21]

In Nigeria oil spillage has hurt, and in some cases destroyed, activity in other economic sectors. Communities in oil producing areas have long called for compensation from the oil companies for damages and recently their anger has led to violence against industry facilities.[22] Some Nigerians warn that oil, which accounts for 80 per cent of the country's foreign exchange, cannot lead to sustainable development. Economic indicators support this view; between 1984 and 1992, Nigeria's foreign debt doubled while its per capita income dropped to the point at which Nigeria now ranks as the world's 20th poorest nation.[23]

Punishing the Whistleblowers

In 1990, a New Jersey federal jury awarded US$1.375 million in damages to Valcar Bowman, a former Mobil environmental affairs manager. Bowman had alleged that Mobil pressured him to alter environmental audit reports and at one point ordered him to remove incriminating documents about air pollution from the Torrance refinery. After he refused, Mobil fired Bowman based on what he calls a "cost-containment sham" in a year of record company profits.[24]

A second former employee, Myron Mehlman, initiated a suit against Mobil in 1990 for wrongful dismissal. Mehlman, who used to be the director of toxicology and manager of the company's environmental health and science laboratories, charges that Mobil fired him in 1989 for calling attention to the dangers of high benzene levels in Mobil gasoline. He also claims that the company incorrectly reported toxicity test results both within Mobil and to outside agencies.[25] Mobil denied the allegations, but Mehlman won the case in 1994 and the company was ordered to pay him US$3.5 million.[26]

Plastic Myths

Mobil claims that "'post-consumer' plastics recycling...has become one of the most important components in the company's environmental agenda."[27] Consumers of plastic, the target of these ads, should be wary of feeling too good about using Mobil plastic, however. A Greenpeace investigation found Mobil plastic bags which had been exported all the way to Indonesia for recycling. The plant manager estimates that up to 40 per cent of all the plastic exported to his plant for "recycling" from the US and Europe is simply dumped (for more about the myth of plastics recycling see Solvay case study).

Endnotes

1. From *Mobil World*, Special Issue, March 1991.
2. For more information on biodegradability and plastics see the Greenpeace report *Breaking Down the Degradable Plastics Scam*, Washington, DC, 1990.
3. Quoted in ibid, p. 2.
4. Quoted in 1990 Corporate Profiles, *Multinational Monitor*, December 1990, p. 14.
5. See "Mobil, FTC to settle 'environmental' claims for its Hefty trash bags," in the *Boston Globe*, 28 July 1992; and Keith Schneider, "Guides on Environmental Ad Claims," *The New York Times*, 29 July 1992.
6. *1991 Mobil Fact Book*, a supplement to the Annual Report, p. 72.
7. *Mobil Annual Report 1991*, pp. 10-11.
8. For more information on the hazards petroleum products pose to the marine environment see Judith Kimmerling with the Natural Resources Defense Council, *Amazon Crude*, 1991, esp. pp. 65-73.
9. Greenpeace, *The Dinosaur's Path The Exxon Valdez, Oil and National Security*, Washington, DC, 1990, p. 9.
10. The information in this paragraph is from Toye Olori, "Nigeria: Petroleum Industry Pollutes the Environment," Inter Press Service International News, 14 February 1992.
11. Council on Economic Priorities (CEP), "Corporate Environment Report: Mobil Oil," New York, 1991.
12. Ibid.

13. Ibid.
14. Quoted in 1989 Corporate Profiles, in *Multinational Monitor*, December 1989, pp. 12-13.
15. CEP, op cit.
16. CEP, op cit.
17. *1991 Mobil Fact Book*, p. 26.
18. CEP, op cit; and 1990 Mobil profile in *Multinational Monitor*.
19. CEP, op cit.
20. CEP, op cit.
21. *The Dinosaur's Path*, p. 9.
22. Olori, op cit., and John Owen-Davies, "Oil Workers Find Life is Harsh in Swamps of Nigeria's Outback Outposts: Employees on Niger Delta endure civil unrest, long hours, little recreation, and lack of female companionship," *The Los Angeles Times*, 12 January 1992.
23. Jato Thompson, "Coping with the debt burden is a major headache," in *African Business*, April 1992, p. 19; and Pini Jason, "Why hasn't six years of SAP revived the economy?," in *African Business*, June 1992, p. 20.
24. 1990 Mobil profile in *Multinational Monitor*; and Ken Sternberg, "Mobil faces employee lawsuit," in *Chemicalweek*, 14 November 1990.
25. Ibid.
26. According to Mobil spokesperson John Lord, November 1994.
27. In *Mobil World*, Special Issue.

Source: *Chemicalweek*, 17 July 1991.

Greenwash Snapshot #3

DOW CHEMICAL COMPANY

A case study in organochlorine contamination, "product stewardship," and avoiding liability.

The Dow Chemical Company (Dow)
Chairman: Frank Popoff
Chief Executive Officer: William Stavropoulos
Headquarters: 2030 Willard H. Dow Center Midland, Michigan 48674, USA
Telephone: 517-636-1000   Fax: 517-636-0922
Major businesses: chemicals; plastics; consumer specialties; hydrocarbons and energy.
Major subsidiary: DowElanco.

Dow operates manufacturing plants in 33 countries; over half of its business is conducted outside the US. Dow is a member of the WBCSD and a signer of Responsible Care and the ICC Rotterdam Charter.

Dow, the US's second largest chemical company and the world's sixth biggest pesticide manufacturer, has spent tens of millions of dollars in advertising to help people forget that it produced DDT, Agent Orange, and ingredients for napalm, and that it has earned a reputation for indifference to public concern and resistance to governmental regulation.[1] Its top executives regularly appear at conferences around the world to display the company's leadership in product stewardship, waste reduction, and other environmental practices.

World Leader in Chlorine Production

> "Our products provide many benefits to society around the world, but
> those contributions are often taken for granted – sometimes even
> questioned. Chlorine, for example...."
> – *from Dow's 1991 Annual Report* [2]

One of Dow's first products was chlorine bleach and the company is the
top manufacturer of chlorine in the world. Dow plants in the US,
Canada, Germany, and Brazil have an annual production capacity of
over 4.1 million tons, or about ten per cent of world production of
chlorine.[3]

Chlorine is involved in about half of all commercial chemistry.[4] Most
chlorine is used either for bleaching pulp or incorporated into
organochlorines to make products such as plastics, pesticides, and
solvents. Organochlorines are ecologically-persistent, bioaccumulative,
highly toxic and are the source of many ecological and health problems
such as ozone depletion, pesticide poisonings, and widespread groundwater
contamination. There is growing opinion that the manufacture, use, and
disposal of chlorine and organochlorines should be phased out worldwide.[5]

Nonetheless, Dow continues to downplay the hazards of chlorine
chemistry, emphasizing only that chlorine is "valuable" for the formulation of products such as pesticides and plastic and for the generation of
caustic soda (see Solvay case study).[6]

Meanwhile, the US government has targeted Dow organochlorines.
In 1989, the US Environmental Protection Agency (EPA) cited seven
Dow facilities as posing a high cancer risk due to toxic air emissions. In
five plants the danger came from chlorinated compounds.[7] The following
year, Dow and other corporate polluters came to an agreement with the
EPA to reduce toxic emissions. Although Dow now cites these reductions
with pride, the 1990 agreement was reached only after long negotiations,
EPA pressure, and were for levels no lower than that which Congress was
expected to impose anyway.[8]

Chlorine and Dioxin

The manufacture, use, and incineration of chlorine and organochlorines form by-products such as dioxins and furans, which are among the most toxic substances known to science. These by-products cause cancer, birth defects, and damage the reproductive, neurological, and digestive systems even at extremely low doses. Dow maintains that dioxins are not harmful to humans.[9]

Dow has resisted governmental regulations to curb dioxin releases. During the mid-1980s, state and federal authorities determined that dioxin discharges from Dow's Midland, Michigan, facility into the nearby Tittabawassee River had contaminated sediment and fish and established limits on future discharges. Dow responded by threatening not to build a new aspirin plant in Michigan (the plant would not produce cancer-causing chemicals). One critic labelled this tactic – using the prospect of economic development and jobs as a lever to ease pollution rules – "classic corporate blackmail."[10] In 1988, an EPA risk management study concluded that Dow was the "most significant" source of dioxin contamination in the Midland area.[11]

Dow facilities have had many accidents involving chlorine or chlorinated compounds. Some examples:

• 1985 – Dow spilled 2500 gallons of a chlorinated cleaning solvent into the St. Clair River in Sarnia, Ontario. The solvent soaked up dioxins from previous spills and formed what the press called a "toxic blob." Dow was fined US$16,000 for the spill and spent another US$1 million in cleanup costs.[12]

• 1986 – Almost 13,000 pounds of chlorinated chemical vapor, including 7,700 pounds of the carcinogen vinyl chloride, leaked into the atmosphere from the Dow plant in Sarnia. It was the sixth toxic chemical accident at the plant in nine months.[13]

• 1989 – 2200 pounds of chlorine gas leaked from a pipe at the Sarnia plant, forcing one worker to go to an on-site health center.[14]

- 1989 – 120 barrels of a Dow-manufactured chlorinated compound en route to the former Soviet Union were damaged in the East Sea, off the coast of Germany, and released toxic fumes that contaminated part of the transport ship and forced six crew members to go to a hospital for observation.[15]

- 1991 – A malfunction at Dow's plant in Pittsburg, California caused 40 pounds of chlorine gas and nearly 900 pounds of carbon tetrachloride gas to escape. Two workers were hospitalized and four others required treatment.[16]

- 1994 – A worker was hurt when 230 pounds of hydrogen chloride gas leaked while being transferred from a rail car into plant equipment at Dow Corning's Midland, Michigan facility. Three hours after the accident, another leak released about 30 more pounds of hydrogen chloride.[17]

ChemAware

Dow formed one of the first product stewardship programs in the industry, in 1972. This program failed to alleviate the problems outlined in this snapshot, a fact tacitly acknowledged by the creation of similar programs with new names in recent years.

In 1989, Dow began to publish and distribute a glossy brochure of what the company calls "the best environmental and Product Stewardship program anywhere," ChemAware. Dow describes ChemAware as an "umbrella program" and boasts about Dow's long-standing commitment to the environment, health and safety, and the company's "legacy" of cradle-to-grave product stewardship.[18] Despite the hyperbole in the brochure and its advertisements, ChemAware is in fact directed towards only a few products and customers relating to the chlorine industry.

Not all of Dow's customers, however, have been satisfied with Dow's product stewardship. In 1988, Charles Gelman of Gelman Sciences sued Dow and other companies for allegedly failing to inform him of the proper disposal of a product they manufactured and marketed. In 1990, Dow filed a motion in court to cancel subpoenas that required depositions from Dow executives. "Apparently Dow and other manufacturers are willing to claim they are responsible for disposal if it helps sales," Gelman asserted, "but deny responsibility if a problem actually develops."[19] The case was settled out of court in 1993.[20]

The Cases of Agent Orange and DBCP

With the help of US federal courts, Dow and other producers of Agent Orange are trying to prevent US Vietnam veterans from bringing another liability suit to trial. Veterans who were exposed to the dioxin-contaminated herbicide have suffered severe health consequences including rare cancers, skin diseases, multiple sclerosis, birth defects in their children, and psychological disorders.[21]

When veterans filed their first class-action suit in New York, the judge used evidence provided by one of the defendants, Monsanto, to deny the veterans a trial in 1984. In 1990, an EPA official accused Monsanto of falsifying its studies and referred the case for criminal investigation (see Monsanto case study).

In 1989, the widow of a Vietnam veteran filed a second class-action suit on behalf of other veterans. Although filed in Texas state court, the federal courts transferred the case back to the original judge, an unusual move that is obviously in the defendants' interest. According to the widow's attorney, "The chemical companies have asked the federal courts – without any legal ground for doing so – to transfer the...case to the Brooklyn federal court just so it could be handled by" the original judge. "They've denied [the plaintiff]...the right to an impartial decision-maker."[22] In April 1992, the judge dismissed the case, which is now being appealed.

Along with Shell Oil, Dow was a co-defendant in the liability case involving the insecticide DBCP and sterility among Costa Rican workers exposed to the it. Like Shell, Dow knew from studies conducted in the late 1950s that DBCP caused sterility in male laboratory animals and failed to include this information on its product labels. Like Shell, Dow pleaded "forum non conveniens" until a Texas court struck down the doctrine in 1990 and allowed the case to proceed (see Shell case study).[23]

"It took us a long time to realize that regulators, legislators, even
environmentalists had a right to ask questions."
  – *Keith McKennon, former President,*
    Dow Chemical USA 1987 [24]

Exporting Unregistered Pesticides

In its 1989 Annual Report, Dow claimed that lower application rates
required for its herbicides Verdict and Gallant were examples of Dow
helping to "safeguard the environment."[25]

Verdict and Gallant are two trade names of haloxyfop, a WHO Class
II "moderately hazardous" herbicide that has never been registered for
use in the United States. In the late 1980s and early 1990s Dow applied
for four product registrations and four food tolerance petitions. Each time
the US EPA refused Dow because of concerns over haloxyfop's
oncogenicity. In 1988, the EPA labelled haloxyfop a "probable" human
carcinogen. Moreover, the formulation of both herbicides includes xylenes,
which are known to damage the liver, kidneys, and bone marrow and are
toxic to the fetus and central nervous system. Despite its dangers, Dow
exports haloxyfop to dozens of countries in Africa, Latin America, Asia,
the Caribbean, and Europe.[26]

Dow manufactures and markets a second pesticide, nuarimol (trade
name Gauntlet), which the EPA refused to register for use in the US, in
this case because of "oncogenicity and teratogenicity [birth defect]
concerns." Dow exports nuarimol to at least 19 countries in Africa, Latin
America, Asia, and Europe.[27]

Dow has had other problems with its agrochemicals. In the mid-
1980s almond growers in Sacramento Valley, California noticed that
crops sprayed with Dow's Lorsban 4E (chemical name chloropyrifos, a
"moderately hazardous" insecticide) were damaged. Dow settled with
some of the growers based on an 18-20 per cent crop yield reduction in
1985. Lorsban 4E is no longer allowed as a dormant spray on almonds in
Sacramento Valley.[28]

In 1989, Colombian authorities ordered Dow Chemical of Cartagèna
to shut down after four tons of Lorsban 4E spilled into Cartagena Bay.
The accident destroyed five tons of fish and threatened the livelihoods of

local fisherfolk. Dow was fined and forced to pay for cleanup costs.[29] Also in 1989, Dow was found to be advertising that Lorsban 4E offers safe use to field staff and applicators, a violation (Article 11.1.8) of the Food and Agricultural Organization's International Code of Conduct on the Distribution and Use of Pesticides.[30]

Dow's 1991 Annual Report Dow boasted"record sales" of Lorsban and Dursban (the trade name for chloropyrifos in Europe.)[31]

Endnotes

1. Jan Loveland, "Dow Chemical's Greenwashing," in *Metro Times* (Detroit), 30 May-5 June 1990; Claudia Deutsch, "Dow Chemical Wants to Be Your Friend," in *The New York Times*, 22 November 1987; and Ross Brockley, "Dow: The Menace from Midland," in *Multinational Monitor*, July/August 1991, pp. 37-40.
2. 1991 Annual Report, p. 9.
3. Dow's plants currently have production capacities of 2,476,056 metric tons/year of chlorine in the US and 710,000 mt/y in Canada. This information available from the Chlorine Institute and Stanford Research Institute. In Germany the figure is 660,000 mt/y and in Brazil, 300,000 mt/y. For overall world production see Joe Thornton, *The Product is the Poison – The Case for a Chlorine Phase-Out*, Greenpeace USA, Washington, DC, 1991, p. 7.
4. "Proposed Chlorine Phase-Out Could Severely Affect Chemical Industry," *Chicago Tribune*, 4 September 1992.
5. For general information on the toxicity of chlorine and phase-out see Thornton, pp. 8-14 & pp. 49-51.
6. 1991 Annual Report, pp. 9-10.
7. United Press International, 8 June 1989.
8. H. Josef Herbert, "Toxic Wastes," Associated Press, 19 September 1990.
9. Brockley, op cit, p. 38.
10. David Waymire, "Dow accused of 'corporate blackmail' over permit," *Bay City*, 2 April 1984. Also US EPA report "Dow Chemical Wastewater Characterization Study, Tittabawassee River Sediments and Native Fish," 15 July 1986, esp. pp. 1-3.
11. Loveland, op cit.
12. David Israelson, "Judge fines Dow $16,000 for toxic blob spill in river," *The*

Toronto Star, 18 February 1986. Also Tom Spears, "Dow Chemical to stop fouling St. Clair River," *Ottawa Citizen,* 19 May 1991.

13. Don Tschirhart, "Toxic vapor escapes near Port Huron," *Detroit News,* 14 May 1986.
14. Canadian Press Newswire, 26 June 1989.
15. Miriam Widman, "Dow Will Remove Toxic Chemicals in West Germany," *Journal of Commerce,* 7 August 1989.
16. "Six Men Injured in Poison Gas Release," *Los Angeles Times,* 7 May 1991.
17. "Two chemical leaks at Dow Corning," UPI, 4 July 1994.
18. See Dow's *ChemAware Enhanced Product and Environmental Stewardship.*
19. PRNewswire, Ann Arbor, Michigan, 1 November 1990.
20. According to Gelman spokesperson Ed Levitt, November 1994.
21. Laura Akgulian, "The Agent Orange Trials," in *Multinational Monitor,* July/August 1991, pp. 20-23.
22. Ibid, p. 23.
23. See: David Weir and Constance Mathiesson, "Will the Circle Be Unbroken," in *Mother Jones,* June 1989, pp. 21-27; Lis Wiehl, "Texas Courts Opened to Foreign Damage Cases," *The New York Times,* 25 May 1990; Patricia Crisafulli, "Costa Ricans File Suit Faulting Pesticides," *Journal of Commerce,* 22 July 1991.
24. Deutsch, op cit.
25. Dow's 1989 Annual Report, p. 17.
26. Sandra Marquardt, Laura Glassman, and Elizabeth Sheldon, *Never-Registered Pesticides: Rejected Toxics Join the 'Circle of Poison',* A Greenpeace Report, February 1992, pp. 22-23.
27. Ibid, pp. 28-29.
28. "Dow's Unsettling Lorsban Settlement," from Outlook, Pesticide Action Network North America Regional Center, 24 April 1991.
29. "Multinational Chemical Plant Closed," *Diario Las Americas,* 29 June 1989.
30. *The FAO Code: Missing Ingredients Prior Informed Consent in the International Code of Conduct on the Distribution and Use of Pesticides,* The Pesticide Trust, London, 1989, p. 66.
31. 1991 Annual Report, p. 27.

Dow Pesticides with Known Dioxin Contamination

Produced in 1995

2-Chloro-4-Phenylphenol
various Chlorophenols
2, 4-D and salts
2, 4-DP and salts
2, 4-Dichlorophenol

Produced in the Past

Chloranil
Erbon
Hexachlorphene
Pentachlorophenol and salts
Ronnel
2,4,5-T and salts
2,4,5-Trichlorophenol
2,3,4,6-Tetrachlorophenol

Dow Industrial Chemicals Associated with Dioxin Generation

Chlorine
Ethylene Dichloride (feedstock for PVC)
Carbon Tetrachloride
Tetrachloroethylene
Trichloroethylene
Epchlorohydrin
Allyl Chloride

– adapted from Appendices 2 and 4 of Jack Weinberg et al, "Dow Brand Dioxin," Greenpeace, Washington D.C., 1995.

protecting
the
skies

A) Cleaning agents – medical devices
B) Toxicity testing at Du Pont's Haskell Laboratory
C) Process development for alternatives
D) Pilot plant for CFC alternatives
E) Alternative refrigerant testing at "Freon" Products Laboratory

DU PONT

Source: From a company brochure on DuPont and Ozone Protection.

Greenwash Snapshot #4

DUPONT DE NEMOURS & COMPANY

A case study in ozone destruction, hazardous exports, and toxic chemical pollution.

DuPont de Nemours & Company (DuPont)
Chairman: Edgar S. Woolard, Jr.
Chief Executive Officer: John A. Krol
Headquarters: 1007 Market St., Wilmington, Delaware 19898 USA
Tel: 302-774-1000 Fax: 302-774-7322
Major businesses: chemicals; petroleum; fibers; polymers; coal; pesticides.
Major subsidiary: Conoco.

DuPont is the largest chemical company in the United States, and has operations in 40 countries. DuPont has signed Responsible Care, the ICC Rotterdam Charter, and is a member of WBCSD.

Hold the Applause[1]

DuPont Chairman Edgar S. Woolard, Jr. has been credited with inventing the phrase "corporate environmentalism." He has referred to himself as "Chief Environmental Officer," saying: "Our continued existence as a leading manufacturer requires that we excel in environmental performance[.]"[2] In the US, DuPont frequently boasts of leadership in health, safety, and environmental practices. A champion of the Rotterdam Charter, Woolard proclaims that the international business community is "...working out a set of principles that will help industry live up to society's expectations around the world."[3]

DuPont's television advertisement, known as "Applause," shows sea lions clapping, ducks flapping, dolphins jumping, flamingos flying, and whales breaching. In the background plays Ode to Joy from Beethoven's Ninth Symphony, while a narrator intones DuPont's announcement that DuPont will "pioneer the use of double-hulled tankers...in order to safeguard the environment." What the ad doesn't tell you: DuPont's double-hulled tankers were not in the water when the ad ran, the full fleet won't be double-hulled until the year 2000, and the sea lions, otters, penguins, and seals depicted in the ad do not live in the Gulf of Mexico where the first tankers will operate.[4]

DuPont's commandeering of images of beautiful wildlife and sounds of rousing music in a desperate attempt to create an impression that they are environmentalists is understandable. The company is a world leader in ozone destruction, has been one of the largest generators of hazardous waste in the US, and was one of the last producers of toxic lead gasoline additives in the world.[5]

World Leader in Ozone Destruction

DuPont invented chlorofluorocarbons (CFCs) – the primary chemicals responsible for ozone depletion – and has in the past accounted for as much as 25 per cent of the global CFC market. Stratospheric ozone protects life on earth from harmful ultraviolet radiation. The United Nations Environment Programme (UNEP) conservatively estimates that the current level of ozone depletion will cause at least 300,000 additional cases of skin cancer, including malignant melanomas, and about 1.5 million additional cases of cataracts annually. Human immune system suppression, damage to crops, and decreases in the phytoplankton population at the base of the marine food chain are also highly likely.[6]

A look at DuPont's history with the CFC and ozone issues is the most telling evidence of the company's greenwash efforts:

1928 – DuPont/GM scientists invent CFCs.
1974 – Scientists link CFCs to ozone destruction. DuPont pledges to stop production if proof is found.
1975 – White House Task Force finds CFCs "cause for concern." DuPont warns against "acting without the facts."
1978 – US Environmental Protection Agency (EPA) and Food and

Drug Administration (FDA) ban non-essential CFC aerosols. DuPont continues selling CFCs for aerosols abroad.

1979 – National Academy of Sciences warns that continues CFC use will lead to 16.5 per cent ozone loss. DuPont says: "All ozone depletion figure to date are based on a series of uncertain projections."

1980 – DuPont takes lead in forming the Alliance for Responsible CFC Policy, an organization which has helped stall a CFC phase-out.

1981 – NASA satellites confirm ozone decline. DuPont discontinues most research on CFC alternatives.

1985 – Scientists discover ozone hole over Antarctica. DuPont expands CFC production in Japan.

1987 – Scientists confirm CFC role in Antarctic ozone depletion. The Montreal Protocol cuts CFC production by 50 per cent (see box). DuPont says: "We believe there is no imminent crisis that demands unilateral regulation."

1988 – Scientists report ozone depletion over temperate zones. DuPont announces phase-out of fully halogenated CFCs, but without a firm timeline.

1989 – Ozone damage over Arctic reported. Helsinki Declaration strengthens Montreal Protocol and orders phase-out of CFCs by 2000. Hoechst Company announces 1995 target date for unchlorinated CFC substitutes. DuPont lobbies against faster phase-out of CFCs.

1990 – Ozone hole opens over Antarctica for 12th straight year. Ninety three nations agree to strengthen Montreal Protocol. DuPont receives "stratospheric protection award" from US EPA.

1991 – With new data, US EPA projects 200,000 additional skin cancer deaths and 12 million skin cancers over 50 years from increased ultraviolet radiation. DuPont blocks shareholder resolution calling for phase-out by 1995.

1992 – Inevitability of northern hemisphere ozone hole confirmed. DuPont buys full page ad in the *New York Times* saying "we will stop selling CFCs as soon as possible," but only "in the

US and other developed countries."[7]

1993 – Measurements show that global levels of ozone hit a 14-year low; the data indicate that the amount of ozone in the northern hemisphere is less than scientists had predicted.[8]

1995 – Scientists who first linked CFCs to ozone destruction win Nobel Prize in Chemistry. According to the UN World Meteorological Organization, the hole in the ozone shield over the Antarctic covered an area twice the size of Europe at its seasonal peak and grew at an unprecedented rate in 1995.
        – adapted from "Hold the Applause"[9]

The Failure of the Montreal Protocol

Although initially hailed as a landmark in environmental protection, the Montreal Protocol is in fact industry-led and does not protect the ozone layer adequately from further deterioration. Because of chemical industry influence, the agreement includes loopholes for the continued production of CFCs and other ozone destroyers for decades to come. The Montreal Protocol even sanctions the use of ozone destroying substances as CFC replacements. In addition, the agreement lacks adequate provisions for funding and technology transfer to less-industrialized countries.

DuPont's Latest CFC Greenwash: "Phase-out," HCFCs, and HFCs

In March 1993, DuPont announced it would stop CFC production for sale in "developed countries" on 31 December 1994. The announcement followed reports from the United Nations World Meteorological Organization and Environment Canada of record-low ozone levels over Northern Europe and Canada.[10]

This does not mean, however, that DuPont ceased making CFCs altogether at the end of 1994. The company continues to manufacture CFCs in the US and other industrialized countries for export to the less-industrialized world until 2010. Moreover, the "phase-out" under the Montreal Protocol includes a loophole for continued CFC production for use in industrialized countries after 1994 based on so-called "essential uses" which will be defined through a petitioning process by CFC producers and users. Given the history of the CFC issue and the role of the chemistry industry, it is likely that "essential uses" will be defined widely.

DuPont has advertised and marketed its substitutes for CFCs – hydrochlorofluorocarbons (HCFCs) and hydrofluorocarbons (HFCs) – as "environmentally enlightened." But these gases also threaten the global atmosphere. HFCs are potent greenhouse gases and are likely to be limited soon either in the US or internationally under a global Climate Treaty.[11] HCFCs are ozone depleters as well as greenhouse gases. According to one researcher, HCFCs are three to five times as destructive to ozone as originally claimed by DuPont.[12] DuPont will produce HCFCs for use in industrialized countries until 2030; no end date has been assigned for sale to the less-industrialized world.

DuPont ignores the existence of alternatives to CFCs which are environmentally safer than HCFCs and HFCs, although such alternatives exist for nearly every CFC application: refrigeration; electronics cleaning; aerosols; foam production; and firefighting techniques.[13] Furthermore, DuPont's promotion and marketing of HCFCs and HFCs inhibit the commercial viability of these safer alternatives.

> "If you have an alternative [to HCFCs and HFCs] that will work, I will support it. I will advertise it and I will buy one for DuPont tomorrow."
> – *DuPont Chairman & former CEO Edgar Woolard, Jr., 1992* [14]

DuPont and Leaded Gas

There is a clear, direct link between environmental lead contamination and brain damage, especially in children. Even low levels of blood lead can cause decreased intelligence, learning disabilities, reduced memory, and behavioral disturbances.

DuPont is the inventor of tetraethyl lead (TEL) gasoline additive, a product which has been virtually eliminated from use in the US, Canada, Japan, and Australia because of its well-known contribution to environmental lead contamination and to childhood lead poisoning. Despite the restrictions on lead additive in the North, DuPont continued to produce and export TEL from New Jersey until 1991 – the site was the last TEL production facility in the US – when declining sales and a stricter lead discharge permit forced an end to production. And until November 1992, DuPont owned 40 per cent of TEMSA, a TEL producer in Coatzalcoalcos, Mexico.[15] Mexico has had one of the worst lead contamination problems in the world.

"Lead in Petrol: The Mistake of the 20th Century."
– *Dr. Carl Shy,*
  World Health Organization, 1990[16]

"Today we have one gasoline for the rich countries, and another deadlier gasoline for less-industrialized countries."
– *Dr. Mario Epelman,*
  Greenpeace Argentina, 1991[17]

"Because of the narrow streets and overcrowding in urban areas, because of the prevalence of dusts both indoor and outdoor, because of poor nutrition and health, poor hygienic conditions and the preponderance of pregnant women and children, the populations of developing countries are much more susceptible to the hazards of environmental lead contamination."
– *Dr. Jerome Nriagu,*
  Nigerian scientist, 1991[18]

"We can expect lead toxicity to be truly epidemic among children in urban centers in many countries in the Third World."
– *Professor David Schwartzman,*
  Howard University, 1991[19]

"If we thought [TEL] was a hazard, we wouldn't export it. We don't export hazards."
– *Dr. Carl Hutter,*
  DuPont Product Manager, 1991[20]

DuPont and the Benlate Scandal

According to farmers in Florida, DuPont's fungicide Benlate DF (chemical name benomyl) – a product with annual sales of US$100 million – has ruined crops and soil. "We have 1,150 acres that's not worth a damn to anybody," lamented one farmer in 1992.[21] Damage to crop and nursery plants after Benlate DF use has also been reported in Costa Rica, Jamaica, Thailand, and the Philippines. DuPont recalled Benlate DF from the market in early 1991, and has also withdrawn the chemical's use in all of Central America.[22]

Although DuPont insisted publicly that there was no proof of Benlate DF's harm, the company paid at least US$500 million to settle lawsuits claiming Benlate DF damage to crops in 40 states, primarily Florida.[23] The company said it made these payments because it was "the moral thing to do."[24]

In November 1992, DuPont stopped settlement payments because it stated that company tests had shown that the fungicide was not the cause of environmental damage. This assertion prompted Florida's Commissioner of Agriculture to respond in a letter to Chairman and then CEO Woolard: "My scientific staff have reached the conclusion that the DuPont company has known for years about certain phytotoxic effects of Benlate.... I am deeply disturbed that your company did not make all information available to state and federal regulators as soon as you learned of it, that DuPont continues to withhold relevant data and that you have failed to live up to your financial responsibilities to Florida growers."[25]

Despite its denial of Benlate's alleged harm, DuPont reversed itself in 1993 and began settling lawsuits again, beginning with one case in August for US$4.25 million.[26] In September 1993, a federal jury in Arkansas found DuPont liable for US$10.7 million in the first Benlate crop damage lawsuit to be decided by jury (DuPont said it would appeal the ruling).[27] And in May 1994, it was announced that DuPont had settled 220 suits for US$214 million, leaving the company with 260 more cases to face. By that time, settlements had cost DuPont US$755 million.[28] "Our wish is not to settle," explained a DuPont spokesperson with a different rationale than the company had used earlier, "but we realize that it is costing more to litigate. So it is strictly an economic decision."[29]

DuPont's legal woes over Benlate have extended to the Caribbean. In November 1994, five dozen farmers from Puerto Rico and the Dominican Republic filed a US$520 million lawsuit alleging that Benlate 50DF had destroyed many of their crops since 1989. DuPont denies the allegations.[30]

> "Benlate has caused as much damage as any natural disaster, and
> DuPont cannot walk away from its responsibility."
> – Bob Crawford, Florida Commissioner of
>   Agriculture, 5 November 1992[31]

DuPont in Goa

TNCs' efforts to avoid accountability can extend to production facilities as well as products. In its negotiations for a nylon plant in the Indian state of Goa, DuPont had drawn up a contract which specifically exonerated the company from any liability which may arise from the facility (this, following the litigation against Union Carbide for Bhopal).[32] The contract and the plant itself were vehemently opposed by the community for over ten years. In October 1994, villagers angry over the government's refusal to reject the project stormed the DuPont site, destroying buildings and burning equipment.[33]

In January 1995, a protestor was killed by police while participating in a blockade of the road to the DuPont site. In June 1995, DuPont announced it was quitting Goa altogether, and would attempt to build the nylon plant in the state of Tamil Nadu, near Madras.

DuPont and Hazardous Wastes

Despite its vaunted environmental leadership, DuPont remains one of the largest producers of toxic wastes on the globe. In 1991, DuPont facilities released over 243 million pounds of toxic chemical emissions in the US – more than any other company according to the EPA (in 1992, DuPont's reported emissions were 231 million pounds).[34] Worldwide, it is estimated that DuPont and its subsidiaries are discharging more than 1.1 million pounds of pollutants every day, or over 414 million pounds per year (this estimate is based on extrapolations of DuPont's 1991 reported releases in the US, however, and is probably conservative).[35]

DuPont itself has refused to provide data about toxic releases from its non-US operations. The company's double-standard behavior suggests that people in other countries are not entitled to the same information about hazardous pollution as those in the United States.[36]

But it is known that pollution from DuPont facilities has affected communities outside the US. In March 1991, the area around DuPont's Quimica Fluor plant in Matamoros, Mexico was deemed so toxic that the Mexican President ordered 30,000 people to give up their homes in order to create a two mile buffer zone around the site. Quimica Fluor has paid US$2.16 million to nearby farmers whose crops were damaged by toxic releases.[37]

> "It's easy to talk a green line these days; it's still hard to walk one."
> – *DuPont Chairman & former CEO Edgar Woolard, Jr., 1991*[38]

Endnotes

1. From Jack Doyle, Friends of the Earth (FOE), *Hold the Applause! A Case Study of Corporate Environmentalism*, Washington, DC, 1991.
2. Remarks of E.S. Woolard, Jr. to the American Chamber of Commerce (UK), London. 4 May 1989, in Doyle, ibid, pp. 3-4.
3. Quoted in David Sarokin, "The Public Data Project, Toxic Releases From Multinational Corporations: Does the Public Have a Right to Know?," Washington, DC, 1992, p. 18.
4. Doyle, op cit, p. 1 and p. 5-10.
5. Doyle, op cit, p. 11, p. 37.
6. United Nations Environment Programme, "Environmental Effects of Ozone Depletion, Executive Summary," November 1991. See also: Kripke, "Effects of Ultraviolet B Radiation on Immune Responses of Mice and Men," testimony before the Senate Commerce Committee, 15 November 1991; and Teramura, testimony before Senate Commerce Committee, 15 November 1991. For DuPont's past market share of CFCs, see Doyle, op cit, p. 37.

7. *The New York Times,* 27 April 1992, p. A5.

8. Vincent Kiernan, "Atmospheric ozone hits a new low," *New Scientist,* 1 May 1993, p. 8.

9. Chronology adapted from Doyle, op cit. See also "Ozone Decay in '95 Is Unparalleled," *The New York Times,* 29 November 1995.

10. Scott McMurray, "DuPont to Speed Up Phaseout of CFCs as Ozone Readings Post Record Lows," *The Wall Street Journal,* 9 March 1993, and Elisabeth Kirschner, "DuPont prepares to speed up chlorofluorocarbon phaseout," *Chemicalweek,* 17 March 1993.

11. Greenpeace, "Climbing Out of the Ozone Hole – Supplement Sources of Safer Alternatives to Ozone Depleting Chemicals by Use Sector and State by State in the US," Washington, DC, January 1993, p. iii.

12. Dr. Arjun Makhijani, "Saving Our Skins: The Causes and Consequences of Ozone Depletion and Policies for Its Restoration and Protection," Institute for Energy and Environmental Research, Tacoma Park, Maryland, 1992.

13. For further information see Greenpeace International, Eds. Sheldon Cohen & Alan Pickaver, "Climbing Out of the Ozone Hole: A Preliminary Survey of Alternatives to Ozone-Depleting Chemicals," Amsterdam, October 1992, passim.

14. Quoted at DuPont's Annual General Meeting, 29 April 1992.

15. From personal communication with DuPont representative, May 1993.

16. Dr. Carl M. Shy, "Lead in Petrol: The Mistake of the 20th Century," in *World Health Statistics Quarterly,* World Health Organization, Geneva. Vol. 43 No. 3, 1990, p. 168.

17. Quoted in Kenny Bruno, "Not Getting the Lead Out," in *Greenpeace Magazine,* October/November/December 1991, p. 18

18. Ibid, p. 18

19. Ibid, p. 18.

20. Christopher Scanlan, "US Firms Exporting Lead That's Banned Here," *Philadelphia Inquirer,* 16 June 1991, in Doyle, op cit, p. 35.

21. Quoted in Anne Raver, "Farmers Worried as a Chemical Friend Turns Foe," in *The New York Times,* 24 February 1992.

22. Kerry Dressler, "Benlate Questions and Secrecy Persist," in "Global Pesticide Campaigner," February 1993, p. 1. Also Elisabeth Kirschner, "Woolard ordered to Benlate hearing, DuPont fined $500,000," in *Chemicalweek,* 19 May 1993, p. 9.

23. "DuPont: Human-Made Agricultural Disaster," in *Multinational Monitor,* December 1993, pp. 11-12.

24. The International Brotherhood of DuPont Workers, "A Warning About the DuPont Company," Martinsville, Virginia, 1992, p. 7.

25. Quoted in "DuPont knew of Benlate threat, state regulator says," UPI, 3 June 1993.

26. Elizabeth Kirschner, "DuPont settles 'first, important' Benlate trial for $4.25 million," in *Chemicalweek*, 18 August 1993, p. 9.

27. Elizabeth Kirschner, "DuPont loses first Benlate jury decision," in *Chemicalweek*, 15 September 1993, p. 6.

28. Allison Lucas, "DuPont settles another round," in *Chemicalweek*, 4 May 1994, p. 10.

29. Ibid.

30. "DuPont faces suit in Puerto Rico," Reuters, 2 November 1994.

31. Quoted in Dressler, op cit, p. 2.

32. Claude Alvarez, ed., *Unwanted Guest – Goans vs. DuPont,* The Other India Press, Goa, India, 1991, p. 4.

33. "Goa villagers' wrath casts shadow over Thapar Du Pont nylon factory," *The Sunday Observer* (India), 16 October 1994. Gary Cohen and Satinath Sarangi, "DuPont Spinning Its Wheels in Goa," in *Multinational Monitor*, March 1995.

34. "DuPont is biggest U.S. polluter in 1991," Reuters, 25 May 1993. Also Allison Lucas, "DuPont: Fighting its Way to a Higher Standard," in *Chemicalweek*, 1 June 1994, p. 41.

35. This estimate based on similar extrapolation made by Doyle, op cit, pp. 19 & 99.

36. Sarokin, op cit, p. 18.

37. Dale Dallabrida, "Mexicans Air Concerns to Du Pont," *Wilmington News Journal,* 7 June 1991, in Doyle, op cit, p. 60.

38. Remarks of E.S. Woolard, Jr. to the Society of the Chemical Industry, Monte Carlo, 8 October 1990, in Doyle, op cit, p. 5.

Source: *Chemicalweek*, 9 December 1992.

Greenwash Snapshot #5

SOLVAY & CIE S.A.

A case study in the greenwash myths of incineration, recycling, and export for recycling.

Solvay & CIE S.A. (Solvay)
Chairman: Baron Daniel Janssen
Headquarters: rue du Prince Albert 33, B-1050 Brussels, Belgium
Tel: 32-2-509-6111 Fax: 32-2-509-6311(Export)
Major businesses: alkalis; plastics (including polyvinyl chloride).

Solvay has operations in over two dozen countries and is the world's fifth largest chlorine and fourth largest polyvinyl chloride manufacturer. Solvay is a signer of Responsible Care and the ICC Rotterdam Charter.

The Myths of Incineration

"During 1990, we placed increased importance on recycling of waste products in our factories. A unit for burning chlorinated waste products yielding hydrochloric acid and energy, has been operating successfully for over a year...It has also been shown that the salts produced in neutralizing the smoke from incinerating domestic waste, can be recycled; as a result, these incinerators have become a viable, non-polluting option."
– *Solvay Report* [1]

That short excerpt, from a section called "Protecting the Future," in Solvay's 1990 annual report, is rich in greenwash myths.

The practice of "recycling" chlorinated wastes which Solvay describes in such glowing terms is in fact a dangerous form of waste incineration. While Solvay boasts that these "units [yield] hydrochloric acid and energy," they neglect to tell the reader that burning chlorinated wastes will also yield new chlorinated compounds, including dioxins, furans, and countless others, some of them more toxic than the original wastes. In the US, the Environmental Protection Agency has identified incinerators which burn chlorinated wastes as the largest known producers of dioxin.[2]

As for yielding energy, Solvay's main businesses – the production of caustic soda and polyvinyl chloride (PVC) for plastic manufacture – have nothing to do with recycling, but they are huge consumers of energy. Heat from waste incineration is captured to supply that energy demand, but the heat recovered from burning is a tiny fraction of the resources contained in many materials. Incineration usually wastes more energy than it recovers. In the end, the process creates air pollution and toxic ash while permanently destroying materials, some of which could be safely recycled or reused.

Solvay's claim that incinerators have become "a non-polluting option" is sheer nonsense; to ascribe this miracle to the fact that salts, which are a small fraction of the incinerator effluent, can be recycled, is ridiculous. Finally, it should be noted that any "recycling" of salts from incinerators is for the purpose of using them in the dirty process of chlor-alkali production.

The Myths of Plastic Recycling

Perhaps no other word has been as abused by corporate environmentalists as "recycling." To recycle – to use again and again – is understandably popular: when properly done, it reduces demand for raw materials, lessens pollution and waste, and saves space and money. It holds an important place in the movement for environmental awareness since it is often the single easiest thing the average person can do to avoid creating environmental problems.

But dirty industries, especially the plastics and waste disposal industries, have appropriated the language of recycling by applying it to processes which do not in any way accomplish the true purposes of recycling. They have taken public enthusiasm for recycling and used it to cover dirty operations. This form of greenwash is one of the most common.

Solvay boasts that its group "has found many new ways of encouraging the recovery of plastic...the material is reused in the manufacture of other products such as pipes, fencing and security barriers."[3]

Solvay's scheme for plastic recycling deserves a hard look before receiving a green stamp of approval. The most important purpose of recycling is to reduce demand for raw materials. But recycling plastic usually involves conversion of used plastic into a totally new product, such as the pipes and fences mentioned by Solvay, and also plastic park benches, plastic flowerpots, and even disposable diapers. This form of recycling actually creates new markets for plastic products and for plastic waste and does nothing to reduce demand for virgin plastic.

Behind Solvay's apparent enthusiasm for plastic recycling lies an unpleasant economic reality. A document leaked to Greenpeace reveals that Solvay's PVC recycling is not economic.[4] While the company stressed the importance of plastic recycling programs to consumers for public relations purposes, Solvay dumped plastic bottles that it collected from the public in containers labeled "Recycle," together with its own PVC wastes, in a landfill in Jemeppe-sur-Sambre in Belgium. Nonetheless, the company got an environmental award for recycling plastic bottles.

The Myths of Export for Recycling

Solvay has also "recycled" waste by sending it abroad. In the 1980s, at least 26 companies including Solvay, ICI, and Bayer sent mercury wastes to Almaden, Spain for eventual recycling. The mercury recycling plant was never built, and the wastes were eventually buried in central Spain, next to a bird sanctuary.[5]

Many operations which export materials for recycling are a cover for waste producers. Most common are schemes to send wastes from wealthy countries to relatively poor countries, where the pollution from dirty recycling operations is unknown or unmeasured. Back home, the company may take credit for its environmental initiative,or even for "waste reduction."

Export for recycling has been one of the major loopholes in agreements to limit the international trade in wastes, and one of the stumbling blocks to enforcing agreements to stop such trade.

Chlorine: The Product is the Poison

Solvay Chairman Daniel Janssen has also been chairman of the European Council of Chemical Manufacturers' Federations (CEFIC), whose subcommittee – the Euro Chlor Federation – works to promote the advantages of chlorine.

Chlorine is at the root of many environmental problems. Ozone-depleting CFCs, pesticides such as DDT and PCP, banned industrial chemicals like PCBs, groundwater contaminants like chlorinated solvents, by-products including dioxins and furans, and literally thousands of other hazardous chemicals get their toxic and persistent qualities from the presence of chlorine.[1]

When combined with hydrocarbons, chlorine produces a class of chemicals called organochlorines. Organochlorines tend to be toxic, persistent, and bioaccumulative; some of the most toxic and persistent organochlorines include dioxins, furans, PCBs, and hexachlorobenzene. In the environment, organochlorines can concentrate to hazardous levels, eventually affecting ecosystems on a broad scale. In humans, they can cause reproductive failure, birth defects, impaired fetal and childhood development, cancer, and neurological damage.[2]

From an environmental point of view, the ultimate goal of all countries should be the phase-out of chlorinated chemicals. But from industry's perspective, a decline in chlorine production is difficult to achieve. This is because chlorine is a co-product with caustic soda, a compound for which demand is strong. As long as this remains the case, chlor-alkali producers such as Solvay will need to sell chlorine. Recently, however, pressure from environmentalists has caused a decline in chlorine demand in the industrialized world as uses in CFCs, pulp bleaching, solvents, and pesticides are phased out.

To compensate for these serious market losses, industry is responding by shifting new-plant chlorine production to less-industrialized regions. Analysts believe production will rise in Latin America, the Middle East, Africa, and especially in Asia. Demand for chlorine is expected to rise in these areas as well.[3]

Whether or not this expectation is fulfilled depends significantly on industry's ability to promote the use of PVC. Chlor-alkali companies like Solvay must look for a "sink" for chlorine, lest it become a waste. PVC – the only major use for chlorine which is not shrinking – is the answer, and it now uses one-quarter to one-third of worldwide chlorine production.[4] Companies consider the less-industrialized world as key to PVC's success; analysts predict that the primary areas of PVC market rise during the 1990s will be Latin America, Asia/Pacific, and the Mideast/Africa, with North America, Japan, and Western Europe trailing behind.[5]

In part, low growth in the industrialized world is due to high levels of consumption already reached in those regions. But PVC increasingly is coming under attack. In Germany, over 60 towns and local authorities have a phase-out program for PVC use in public buildings. The Swedish company IKEA, one of the world's largest furniture distributors, announced in 1991 that it would use environmentally sound substitutes for PVC. Sweden's parliament has endorsed a nationwide plan to phase out all PVC use. And Irma, Denmark's biggest supermarket chain, has achieved a 99 per cent reduction of PVC.[6]

In the new-growth regions of the South, manufacturers peddle PVC as ecologically positive and a part of an improved standard of living. The trade journal *Modern Plastics* is bullish on all-PVC houses: it reported that a Canadian company is marketing them in Mexico, Thailand, and Central and Eastern Europe, invoking the need for economical housing.[7] In Thailand, another company advertises PVC as a "green" substitute material for wood products. Using the slogan "Save the Tree – Use PVC," this company suggests that greater PVC use will help prevent deforestation.[8]

The central reason behind the recent proliferation of frivolous and wasteful PVC packaging and the promotion of a market for PVC building materials has nothing to do with environmentalism. The chlor-alkali industry is not interested in reducing PVC demand through recycling or sensible materials policies. Projects which are touted as important investments or as "sustainable development" may in fact be a sophisticated form of chlorine dumping.

Solvay is one of many chlor-alkali companies which have recently expressed interest in Central and Eastern Europe. Their proposals there should be examined with a skeptical eye so that the greenwash myths they have promoted do not facilitate the creation of yet another waste dump or incinerator masquerading as "sustainable development."

Endnotes

1. From Solvay's 1990 Annual Report, under section "The Environment: Protecting the Future."
2. Joe Thornton, Greenpeace, *Achieving Zero Dioxin – An Emergency Strategy for Dioxin Elimination*, Washington, DC, July 1994, p. 28.
3. From Solvay's 1990 Annual Report.
4. The Solvay document is "Proces-Verbal de la Seance Conseil D'Enterprise," 3 April 1990.
5. See Greenpeace report *Mercury Wastes at Almaden*, Waste Trade Scandals, 1991, p. 3 and passim.

Notes in box on page 96

1. See Joe Thornton, *The Product is the Poison – The Case for a Chlorine Phase-Out,* Greenpeace USA, Washington, DC, 1991, especially p. 1-14.
2. Ibid.
3. Michael Roberts, "Structural Changes in Chlor-Alkali – Shifting Geography and Markets," in *Chemicalweek,* 4 November 1992, p. 36.
4. For information on PVC, see Alasdair Nisbet, "Chlorine and Caustic Soda," in *European Business*, October 1989, pp. 48-49. See also *Chemical Business,* December 1989, p. 25.
5. Roberts, op cit.
6. PVC – Toxic Waste in Disguise, Section 10, "PVC Bans and Phase-outs," Greenpeace International, Amsterdam, 1992.
7. "Plastiscope – Construction – Worldwide market emerging for exportable housing," *Modern Plastics,* v. 69, no. 8, 1 August 1992, p. 21.
8. From advertisement in *The Bangkok Post*, 17 February 1991.

Chlorine Facts

Number of stratospheric ozone molecules that a single chlorine
   radical can destroy: 100,000

Number of years ozone depletion will continue even if all CFC
   production were stopped today: 100

Number of new skin cancers worldwide expected each year due to
   ozone depletion: 300,000

Countries experiencing decrease in chlorine use for pulp bleaching:
   US, Canada, EU, Scandinavian countries

Countries experiencing increase in chlorine use for pulp bleaching:
   Brazil, Indonesia, India, China, Chile

Per cent of US citizens with measurable levels of chlorinated
dioxins in their tissues: 100

Number of organochlorines identified in tissues and fluids of US
   and Canadian population: 177

Per cent of chlorine associated with dioxin during life cycle: 100

Chlorinated waste burned in US hazardous waste incinerators, as
   per cent of all wastes burned: 40

Largest group of dioxin sources in the US: waste incinerators

Product causing most dioxin during life cycle: PVC

Country aiming to phase out all PVC use: Sweden

Continent with low or no growth PVC markets: Europe

Continent with high growth PVC markets: Asia and South America

Percentage of chlorine now used for US PVC manufacture: 35

Projected percentage of chlorine used for PVC manufacture in
2005: 48

Percentage of chlorine used for pulp bleaching: 18

Percentage of chlorine used for water disinfection: 1

Estimate of number of bladder and rectal cancers in the US caused
   each year by chlorination of drinking water: 9,700

Percentage of Persistent Organic Pollutants prioritized for phase
   out by the United Nations which contain chlorine: 100

— adapted from "Just the Facts," compiled by Jay Palter, Greenpeace Canada

Paraquat and Nature working in perfect Harmony

Its a FACT. PARAQUAT is Environmentally friendly. For over 30 years Nature and PARAQUAT have been working in perfect harmony. PARAQUAT's unique formula provides continued variety and abundance of food without upsetting nature's delicate balance.

Ground water, rivers, streams and lakes are not affected by PARAQUAT. Because of PARAQUAT'S adsorption to minerals and clay particles in the soil, it is deactivated rapidly and cannot be released to contaminate ground water and waterways. Once adsorbed, it is biologically unavailable and cannot be taken up by plant roots, earth worms and other organisms. It also causes no adverse effects to fish and other forms of aquatic life.

PARAQUAT is not harmful to our Wildlife. For many decades, PARAQUAT has been used to produce crops for Malaysia, causing no adverse effects on our wildlife.

Soil structure is unaffected by continuous and extended use of PARAQUAT. PARAQUAT DOES NOT destroy the root system. As such, it prevents soil erosion, helps retain moisture in the soil and contributes to increased organic matter in the soil and improves soil tilth and fertility.

PARAQUAT Since its development in 1954, it has withstood the test of time and the most intensive scrutiny by Scientists and numerous Environmental Protection Agencies. PARAQUAT is a unique chemical that has and will continue to be an important tool for food production in Malaysia and around the world.

PARAQUAT For a Greener Malaysia
ICI Agrochemicals

ICI Agrochemicals (Malaysia) Sdn Bhd
Padang Jawa, P.O. Box 185, 41720 Klang.
Selangor Darul Ehsan.

Source: *The Star*, 6 April 1993.

Greenwash Snapshot #6

IMPERIAL CHEMICAL INDUSTRIES PLC & ZENECA GROUP PLC

A case study in pesticide promotion and ozone destruction.

Imperial Chemical Industries PLC (ICI)
Chairman: Sir Ronald Hampel
Headquarters: Imperial Chemical House Millbank, London SW1P 3JF, UK
Tel: 44-171-834-4444 Fax: 44-171-834-2042
Major businesses: paints; explosives; industrial chemicals.

Zeneca Group PLC (Zeneca)
Chairman: David Barnes
Headquarters: 15 Stanhope Gate, London W1Y 6LN, UK
Tel: 44-171-304-5000
Major businesses: pharmaceuticals; agrochemicals; seeds; specialty chemicals.

ICI has manufacturing operations in 40 countries and sales affiliates in 150 countries. One of the largest chemical companies in the world, ICI is a signer of the ICC Rotterdam Charter and Responsible Care. It is a member of the WBCSD.

In 1993 for financial reasons, ICI split off its subsidiary Zeneca, which became a separate holding company for what were ICI's pharmaceuticals, agrochemicals, and specialities businesses. ICI has maintained subsidiaries in India, Pakistan, and Malaysia, however, which produce agrochemicals. ICI's former chairmain Sir Denys Henderson remained as Zeneca's chairman.

For years, ICI (now Zeneca) was the world's biggest producer of paraquat, an extremely toxic herbicide banned in five countries, severely restricted in two others, and unregistered in three more. The Pesticide Action Network has labeled paraquat one of the "Dirty Dozen" pesticides which collectively cause many deaths and widespread environmental damage every year but are still promoted and used heavily in the less-industrialized world.[1]

Paraquat is the world's second largest selling pesticide with annual sales of US$460.[2] In 1987, paraquat accounted for about one-quarter of ICI's total pesticide sales. In 1992, ICI representatives would not reveal what percentage of these sales paraquat represented,but said it was a "significant product" which ICI marketed in over 130 nations.[3]

Although it is not surprising that ICI advertised paraquat aggressively, the lengths to which the company has gone in its promotion can be shocking. In one ad headlined by the caption "Paraquat and Nature Working in Perfect Harmony," ICI's Malaysian subsidiary claimed as "fact" the herbicide's "environmentally friendly" impact on water, land, and wildlife. Interestingly, while it touted paraquat as an "important tool for food production," the ad's aim was essentially defensive: to counter the herbicide's justifiably hazardous reputation and negative image. As such, the ad is an example of pure, unadulterated greenwash.[4]

Paraquat and the Environment

The assertions which ICI's Malaysian subsidiary made about paraquat's environmental effects are so outrageous that they demand point-by-point rebuttal.

ICI Claim #1. *Because paraquat is absorbed onto clay and mineral particles in soil, it is deactivated and cannot be released to contaminate and harm the environment.*

Paraquat, which binds tightly to soil particles, tends to degrade minimally; instead, it stays in the soil. However, according to the US Environmental Protection Agency (EPA), the paraquat degradate QINA (C-carboxy-1-methyl pyridium) is "very loosely adsorbed on the organic matter/clay complex and...has the potential for groundwater contamination."[5]

While undegraded paraquat in soil is difficult to dislodge, this does not mean that it is "deactivated." Rather, because adsorption onto material such as clay helps protect paraquat from breakdown, it increases the herbicide's environmental persistence and accumulation. This in turn enhances paraquat's potential for contamination and damage.

Particles with paraquat can be mobilized by water movement, precipitation, or atmospheric entrainment. These paraquat-particles can be ingested, inhaled, or contacted directly. Enzymes and acidic conditions in the digestive systems of most animals can be expected to release the paraquat. Paraquat on very small particles which have been inhaled can be dissolved in cells within the lungs.[6]

ICI Claim #2: *Paraquat does not affect soil structure adversely, but instead helps improve soil tilth and fertility.*

Research in 1981 showed that paraquat had a negative impact on nitrogen fixing soil rhizobial species associated with alfalfa at all concentrations at which it was tested. In 1984, an Indian researcher observed a powerful mutagenicity effect of paraquat on nitrogen fixing blue-green algae found in rice paddies. Both studies raise serious concerns about the effects of paraquat on nitrogen fertility management from non-chemical sources in some agricultural cropping systems.[7]

ICI Claim #3: *Paraquat cannot harm fish and other aquatic life.*

After spraying with paraquat, an oxygen demand can be created by the cessation of photosynthesis and decay of plants. This can have a harmful effect on fish. Algae has been shown to be especially sensitive to paraquat.[8] Damage to algae might lead to wider aquatic ecosystem disruption, including among fish populations.

Direct sensitivity to paraquat varies among fish species and conditions of application. A report to the US EPA showed that yearling salmon exposed to paraquat in freshwater experienced reduced survival when introduced into saltwater (to which they migrate). A study cited by the World Health Organization indicated that trout exposed to paraquat concentrations over 16 days exhibited a 30 per cent mortality rate. Carp fingerlings exposed to paraquat in the presence of a submerged aquatic plant were more susceptible than those exposed in a weed-free environment.[9]

Paraquat has been found to be fatal to frog tadpoles at the lowest dose tested. Noting that some plants can concentrate paraquat over 2,000 times, researchers raised the concern that consumption of contaminated plants or plant detritus by tadpoles might cause death among such sensitive species.[10]

Claim #4: *Paraquat has no adverse effects on "our [i.e., Malaysian]" wildlife.*

Research indicates that paraquat is highly toxic to mammals as well as some birds and insects. It is difficult to believe that its effects on Malaysian wildlife are benign – unless animals and birds in Malaysia possess some special immunity from paraquat.

US researchers have demonstrated that paraquat kills honeybees at doses lower than those used for weed control and can also be acutely toxic to mite species. Species of deer mice in a field sprayed with paraquat have been found to suffer from liver damage. This study's researchers concluded: "[paraquat] may have long-term effects on wildlife under natural field conditions."[11] Paraquat is extremely toxic to hares. Horses allowed to graze on pastures recently sprayed with paraquat developed lesions in the mouth and had increased mucous secretions. The World Health Organization (WHO) recommends that "all domestic animals should kept far from freshly-sprayed areas."[12]

Paraquat has been shown to cause increased mortality and reduced growth rates in nestlings of the American kestrel. Researchers note the possibility that the birds were poisoned by contact through spraying or by ingestion of contaminated prey such as insects or rodents. A number of US and European studies show that paraquat is also extremely toxic to bird embryos when applied topically to eggs of chicken, mallards, and Japanese quail.[13]

After the Consumers' Association of Penang protested to the Malaysian Pesticides Board, the Board asked ICI's affiliate to withdraw the ad, which subsequently disappeared from local newspapers.[14]

Paraquat and Human Health

Paraquat is acutely toxic to people and can be inhaled, ingested, or absorbed through skin. If swallowed, less than a teaspoon of the herbicide can kill an adult, and there is no antidote. Very low levels of paraquat exposure can lead to skin injuries such as severe dermatitis, burns, and rashes. The US EPA has classified paraquat as a "possible human carcinogen." Epidemiological research in Canada suggests that paraquat may be a factor in the incidence of Parkinson's Disease.[15]

Paraquat use is widespread and poisoning by the herbicide occurs globally. Although lack of monitoring and poor reporting make precise figures impossible to ascertain, paraquat is probably responsible for at least a thousand of deaths annually.[16] Conditions in less-industrialized countries especially can make application of paraquat (and other agrochemicals) a dangerous occupation:

- Illiteracy among agricultural workers prevents them from understanding instructions and warning labels.

- Protective gear and clothing are often unaffordable or unavailable to workers. Moreover, the hot and humid climate in tropical regions renders such clothing unsuitable.

- Access to clean water supplies and medical care may not exist.

- Regulations overseeing pesticide use are often inadequate.

In Costa Rica, a physician reports that he has treated hundreds of farm workers, children, and others poisoned by paraquat."Take it off the market," he says, "It's killing people all over the world."[17]

In Malaysia, paraquat poisoning is common and more than half of such poisonings are fatal. Seventy per cent of these deaths are suicides, the majority of which are committed by agricultural workers who are among the poorest people in the country. The other 30 per cent occur from accidental poisoning or from exposure while handling the herbicide. In a 1988 survey, the Malaysian Department of Agriculture found that only

11 per cent of the workers had been trained to handle paraquat and that
67 per cent of plantation management had failed to provide the workers
with protective clothing. The survey also discovered that almost two-
thirds of the workers reported symptoms of poisoning.[18]

Because paraquat sales have been so profitable for ICI, the company
has fought hard to maintain its markets, even when nations take steps to
protect their people. In 1991, when the President of the Dominican
Republic issued a decree banning 20 hazardous pesticides including
paraquat, ICI sent an emissary to lobby for paraquat and to assure the
Dominicans that his company was very concerned about environmental
matters. ICI also published a multi-colored, promotional brochure for
paraquat. In response, a National Commission for the Environment was
formed and provided a thorough refutation of ICI's claims, detailing
paraquat's many adverse health and environmental effects. Despite the
Presidential decree, paraquat is still sold illegally in the Dominican
Republic.[19]

Product Stewardship for Paraquat: Prevention or Promotion?

To protect its market and image, ICI has attempted to assuage public
concerns about the hazards of paraquat and other pesticides. Besides
adopting the Food and Agricultural Organization's voluntary International Code on the Distribution and Use of Pesticides, the company has
also implemented a "product stewardship" program to develop pesticide
labelling improvements and education tools such as posters and videos.
The professed aim of stewardship is to help prevent products from
harming people or the environment. Encouraging the "safe" application
of chemicals such as paraquat, however, can have the effect of promoting
the product.[20]

ICI's video "The Safe and Effective Use of Gramoxone" (Gramoxone
is a trade name for paraquat) distributed in Malaysia is illustrative. Saying
that it failed to address adequately the dangerous realities of paraquat use,
Sahabat Alam Malaysia (SAM-Friends of the Earth Malaysia) asserted
that the video served "to promote sales of paraquat." "The high acute
hazard and lack of adequate or effective protective equipment and

clothing to eliminate problems associated with paraquat exposure," SAM continued, "cannot be washed away by a glossy film production."[21]

For a product as toxic as paraquat, the only sensible course of stewardship is to plan a global phase-out of its use. This would be a genuinely preventative approach to product stewardship. ICI refused to follow this course and, instead, made plans to build a new US$58 million paraquat manufacturing plant in the UK which will have an annual production capacity of 8,000 tonnes.[22]

Although there are a number of viable non-chemical alternatives to paraquat use, ICI representatives have claimed that pesticides are better for local people's "quality of life" than such options.[23] But as long as ICI, and now Zeneca, continue to make, sell, and profit from paraquat, more paraquat-related poisonings and deaths around the world will doubtless occur.

> "We have a product that's useful, beneficial and which can be managed properly. So why should we take it off [the market]?"
> – *ICI official George Allen, on paraquat, 1991*[24]

ICI and Ozone Depletion

Until recently, ICI was the world's #2 producer of ozone-depleting CFCs. Now the company has begun advertising its leadership in ozone layer protection: "ICI has responded quickly and effectively...ICI is at the forefront...ICI is a world leader...."[25]

In reality, the company failed to acknowledge the link between CFCs and ozone destruction until forced to do so by overwhelming scientific evidence and public opinion, and, like DuPont, its response to the urgent need to stop CFC manufacture has been to continue production while simultaneously developing "drop-in" chemical substitutes which are themselves toxic, ozone-depleters, and greenhouse gases – although less potent than traditional CFCs.[26]

ICI now takes credit for moving away from its CFC businesses. But the net effect of this decision will be less than the company suggests. In 1992, for example, ICI announced that it would stop producing CFCs at its plant in Runcorn, UK. ICI and the Dutch chemical giant Akzo agreed,

however, to concentrate CFC production in the Netherlands beginning in
1993. ICI's customers which are unable to use substitutes will be supplied
by the Akzo plant.[27]

Endnotes

1. Pesticide Action Network, "Global Pesticide Campaigner," v. 1, no. 3, June
    1991, pp. 14 & 16.
2. *Agrow,* no. 178, 19 February 1993, p. 2.
3. According to ICI representative, 1992.
4. The advertisement ran in *The Star*, 6 April 1993, and other Malaysian dailies
    throughout the month of April.
5. From US Environmental Protection Agency data, quoted in Greenpeace
    International & Pesticide Action Network/International Organization of
    Consumer's Unions Penang, Malaysia, "Review of the Pesticide Paraquat
    For Submission to the World Bank Pesticide Advisory Panel," Washington,
    DC, 6-7 December 1989, p. 13.
6. Ibid, and Mary O'Brien, "Paraquat," in *Journal of Pesticide Reform,*
    Summer 1989, p. 29.
7. "Review of Pesticide Paraquat," op cit, p. 13, and O'Brien, op cit, p. 29.
8. "Review of Pesticide Paraquat," op cit, p. 14.
9. "Review of Pesticide Paraquat," op cit, pp. 13-14, and O'Brien, op cit, p. 29.
10. Ibid. O'Brien, op cit, p. 29, and "Review of the Pesticide Paraquat," op cit,
    pp. 15-16.
11. O'Brien, op cit, p. 29.
12. Quoted in "Review of Pesticide Paraquat," p. 15.
13. Ibid, and O'Brien, op cit, pp. 29-30.
14. See letter to the editor of S.M. Mohd Idris, Consumers Association of
    Penang, in the *Multinational Monitor,* July/August 1993.
15. "Review of Pesticide Paraquat," op cit, pp. 3-12, and O'Brien, op cit, pp. 27-
    29.
16. "Review of Pesticide Paraquat," op cit, p. 7.
17. Quoted in Christopher Scanlan, "U.S.-barred products sold abroad," *The
    Miami Herald*, 27 May 1991.
18. Sahabat Alam Malaysia (SAM), "Paraquat in Malaysia: The Death Toll
    Mounts," *Journal of Pesticide Reform,* v. 9, no. 2, Summer 1989, pp. 24-25.
19. Pesticide Action Network, "Dominican Republic Bans Dirty Dozen –
    Industry Fights Back," in "Global Pesticide Campaigner," v. 2, no. 1,

February 1992, pp. 3 & 5.

20. Barbara Dinham, "ICI Limited Paraquat Production--Starting the Chain, A Report for Greenpeace International," London, December 1989, p. 10.

21. SAM Press Statement, "SAM Concerned Over ICI Video on "Safe" Use of Paraquat," 1 August 1987, p. 3. See also SAM's "Comments on ICI's Promotional Paraquat Video 'The Safe and Effective Use of Gramoxone.'"

22. *Agrow,* op cit, and Natasha Alperowicz, "ICI applies for permit to build paraquat plant," *Chemicalweek,* 10 February 1993.

23. According to ICI representative, 1992.

24. Quoted in Christopher Scanlan, "Warning: U.S. exports hazardous to health," *Detroit Free Press,* 19 May 1991.

25. From "ICI and the Environment," Brown Knight & Trustcott, England, pp. 9-10.

26. Council on Economic Priorities, draft environmental profile of ICI, 1991, p. 4.

27. "Akzo and ICI to Pool CFC Production," Reuters, 18 February 1992.

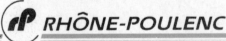
Source: *Chemicalweek*, 17 June 1992.

Greenwash Snapshot #7

RHONE-POULENC S.A.

A case study in toxic pesticide exports and pollution in Brazil's "Valley of Death."

Rhone-Poulenc S.A. (Rhone-Poulenc)

Chairman and Chief Executive Officer: Jean-Rene Fourtou

Headquarters: 25, Quai Paul Doumer, 92408 Courbevoie Cedex, France

Tel: 33-1-47-68-12-34 Fax: 33-1-47-68-16-00

Major businesses: agricultural, industrial, and specialty chemicals and intermediates; pharmaceuticals; fibers.

Major subsidiary: Rhodia (Brazil)

Rhone-Poulenc is France's largest chemical company, world's 8th largest, with 200 operations in over 80 countries. It is 56 per cent owned by the French government. Rhone-Poulenc is a signer of Responsible Care and the ICC Rotterdam Charter. It is a member of the WBCSD.

In 1990, Rhone-Poulenc was named France's "Exporter of the Year" by the French Centre for Foreign Trade. The Centre omitted to mention that many of those exports are hazardous. Given Rhone-Poulenc's pesticide trade practices, it might deserve recognition instead as the "Hazardous Exporter of the Year."[1]

Exporting "The Most Toxic Pesticide Now On Sale"

Rhone-Poulenc is one of only two producers in the world of what has been called "the most toxic pesticide now on sale": aldicarb (trade name Temik).[2] A drop of aldicarb absorbed through the skin can kill an adult. Exposure to smaller amounts can cause nervous disorders, respiratory arrest, nausea, stomach cramps, diarrhea, and headaches.[3] Infants and children are especially vulnerable to poisoning by aldicarb. Aldicarb is also highly toxic to mammals, birds, and aquatic organisms.[4]

In 1985, over 1,000 people in the US and Canada became ill after eating aldicarb-treated watermelons. In 1986, during its first week of use in Costa Rican banana plantations, 100 workers were poisoned by aldicarb. By 1988, Costa Rica's National Institute of Insurance estimated that aldicarb was the number one cause of pesticide poisonings in Guaypil, the country's banana-growing region. Rhone-Poulenc claims that "aldicarb has been used on bananas for more than ten years and has not resulted in any adverse effects."[5]

Aldicarb has contaminated groundwater in at least 27 US states and ranked seventh as a cause of pesticide poisonings requiring hospitalization in California. In 1989, US Environmental Protection Agency (EPA) toxicologists estimated that tens of thousands of infants and children a day were exposed to enough aldicarb residues in bananas and potatoes to pose the risk of illness and recommended that the EPA forbid aldicarb's use on these crops.[6]

Because of its dangers, aldicarb is banned, severely restricted, or unregistered in at least 13 countries.[7] But Rhone-Poulenc continues to sell it in more than 70 nations.[8]

Bhopal Gas in Black America

Rhone-Poulenc has also "exported" the hazards of aldicarb production – not to Southern nations, but to the African-American community of Institute, West Virginia. The plant, bought from Union Carbide in 1986, is now one of the only places where aldicarb is made.[9] Part of the aldicarb manufacturing process includes the use of methyl isocyanate – the gas which killed thousands at Bhopal. Citizens of Institute have said that the aldicarb plant is part of a pattern of corporations locating extremely hazardous facilities in black communities which the residents call "environmental racism."[10]

In 1985, before Rhone-Poulenc owned the Institute plant, a gas leak from the facility sent 134 people to the hospital. Then, residents complained of poor communication and a slow emergency response. But when the Institute plant leaked the Bhopal gas in February 1990, Rhone-Poulenc did not report the leak to the authorities or to the public. In fact, the community did not learn of the leak until a local TV station reported the incident three days after it happened.[11]

> "Good fences rarely make good neighbors. Cooperation always does. We see it happen day after day in community after community ....The first principle of Responsible Care is community outreach – actively seeking to identify local concerns and working with our neighbors to address them."
> – *From Rhone-Poulenc ad, with a picture of smiling employees from the Institute, West Virginia facility*[12]

> "It took a lot to gain our trust. Then they come back and do the same thing again. You can't trust them."
> – *Gail Parker, Institute resident, February 1990*[13]

Lindane

Rhone-Poulenc is one of the world's two major manufacturers of another highly dangerous pesticide: lindane. Lindane is toxic, bioaccumulative, and persistent in the environment; its effects include dizziness, nausea, reproductive disorders, central nervous system damage, and cancer. Lindane has been detected in the breast milk of Argentine women and in rice grains destined for human consumption in Malaysia.[14] Scientists have found lindane and related compounds in coastal waters of the North Sea, in Antarctic lichens and mosses, in seals and foxes in Greenland, and in air and water throughout the world.

Currently, a container with six tons of Rhone-Poulenc's lindane is sitting at the bottom of the English Channel. In 1989, the ship carrying the toxic cargo sank. When the packages of lindane degrade, scientists believe that contamination could severely impact the North Sea ecosystem.[15]

Lindane is banned in 15 countries and severely restricted in ten more.[16] In 1991, the World Bank agreed that the financing of lindane for use in the protection of stored commodities, in public health, and veterinary applications would cease immediately. All additional uses of lindane would be phased out by the Bank by 1993.[17] Despite its health and pollution problems, Rhone-Poulenc continues to market lindane in many countries.

> "The future of agrochemicals hinges on irreproachable ethics."
> — *From Rhone-Poulenc's Annual Report, 1990* [18]

Rhone-Poulenc in the "Valley of Death"

Rhone-Poulenc, like many TNCs, has dirty plants outside its home country. Take Cubatao, Brazil, known as the "Valley of Death" and one of the most polluted cities on earth. It was there, from 1976-78, that Rhone-Poulenc subsidiary Rhodia produced pentachlorophenol (PCP), a hazardous wood preserving chemical. The chemical, known locally as "Chinese powder," was dumped near the community and allegedly caused numerous poisonings among workers and residents.[19] Rhodia was Rhone-Poulenc's most prosperous subsidiary throughout the 1980s, with 11 per cent of its worldwide sales.[20]

It is almost impossible to describe the magnitude of the environmental crisis in Cubatao and its impact on residents. A monitoring device in the Cubatao slum of Vila Parisi has recorded daily doses of 473 tons of carbon dioxide, 182 tons of sulfur, 148 tons of particulate matter, 41 tons of nitrogen oxide, and 31 tons of hydrocarbons.[21] The city has the highest level of acid rain ever recorded.[22]

Emergency air quality alerts are declared in Cubatao dozens of times each year. In July of 1991, the governor of Sao Paulo state suspended industrial activity in Cubatao during a thermal inversion which elevated the particulate content of the air to 2000 parts per cubic meter (66 parts per cubic foot). The international standard for "good quality" air is 50 ppcm, and the norm for a chronically polluted city like Sao Paulo is 70 ppcm.[23] Some residents are exposed each day to around 1200 particulates per cubic meter, more than twice levels that the World Health Organization says provoke "excess mortality."[24]

The city suffers the highest infant death rate in Brazil with one-third of children not surviving their first year. A study released in the 1980s showed that eight per cent of live birth babies suffered from such abnormalities as spinal problems, missing bones, and brain deficiencies.[25] Over one-third of Cubatao's residents suffer from pneumonia, tuberculosis, emphysema, and other respiratory sicknesses.[26]

With dozens of factories in the valley, it is very difficult to identify Rhone-Poulenc's specific contributions to the general, appalling conditions in Cubatao. But even the company has admitted a role in the disaster, saying that "With respect to environment...Cubatao is the Achilles Heel of Rhodia."[27]

In June 1993 a judge from Cubatao ordered the immediate closure of the Rhodia plant. The order followed a report from the Prosecutors Public Service which revealed the existence of huge quantities of PCP and hexachlorobenzene (HCB) contaminated soil at Rhone-Poulenc's facility. Exposure to those organochlorine chemicals can cause liver and nervous system disorders and suppression of the immune system. The company had concealed the contaminated deposits, which were 7000 to 15,000 times higher than legal contamination levels. After sampling the blood of more than 80 per cent of the factory workers, scientists found levels of HCB ranging from six to 16 micrograms/deciliter in all but one of the employees. The judge's decision was made to avoid the continued exposure of plant workers to the toxic residues.[28]

Rhone-Poulenc claims that since 1984 an incinerator at Rhodia's facility, a US$21 million investment, has disposed of all its organochlorine waste products (for the problems with incineration of chlorinated wastes, see Solvay case study).[29]

> "We are considered good citizens. That which is good for us also has to be good for the country."
> – *Rhodia President Edson Vaz Musa, 1986* [30]

> "Borne by the Brazilian [economic] miracle, Rhodia has built an industrial base and powerful market on the American continent..."
> – *Rhone-Poulenc Magazine 'Presence,' 1986* [31]

> "On my knees I address the world. For God's sake, help us! I ask all the presidents of the world; Help us solve our problem. Tell our President

and our public health service. We're slowly dying. We don't see our children grow up....The whole village has been poisoned. But the company knows how to buy silence...I hope everyone will see this. They'll understand why we fight so hard for our families."
— *Francisco Alves de Moura,*
  Cubatao resident, 1987[32]

"...I saw the trucks from Rhodia when we came out of school around eleven at night. They dump all that filthy waste in the forest...We get a northwesterly wind here. It blows very strongly. The wind carries the chemical dust to the village, and we all get itchy....The authorities have just put up a sign: 'Swimming prohibited.' The soil is contaminated. Our trees don't bear fruit anymore."
— *Adauto Alves de Nobrega,*
  Cubatao resident, 1987[33]

"We are paying for the country's development with our lives."
— *Jose de Santana,*
  Cubatao resident, 1985[34]

Rhone-Poulenc's Greenwash Fable

Phosphate pollution, mostly stemming from detergent use, is a major cause of environmental degradation of lakes, rivers, and seas. When phosphate use restrictions in Germany, Switzerland, and Italy directly threatened Rhone-Poulenc's sales, the company hired the public relations firm Hill and Knowlton to mount a campaign to thwart regulations on phosphates in France.[35] The ad campaign, featuring a wolf in sheep's clothing peddling phosphate-free detergent, tells you that "phosphates, from the point of view of...environmental impact, give the best results."

The campaign worked. In 1991, the French government suspended legislation against phosphate use. Who is the real wolf in sheep's clothing?

Endnotes

1. Rhone-Poulenc Annual Report 1990, p. 18.
2. According to Professor Francois Ramade of the University of Orsay, in "Des pesticides aux armes chimique," *La Recherche,* March 1990. For information on aldicarb (and other pesticide) manufacturers, see *Farm Chemicals Handbook '94,* v. 80, Meister Publishing company, Willoughby, Ohio.
3. Pesticide Action Network (PAN) Update Service, "Dangerous Levels of Aldicarb Found in Bananas," September 1991. See also Kai Siedenburg, "High Levels of Aldicarb Found in Bananas," in "Global Pesticide Campaigner," June 1991, pp. 10-11.
4. Siedenburg, ibid, p. 11.
5. Siedenburg, op cit, p. 10.
6. PAN Update Service and Siedenburg, op cit.
7. Demise of the Dirty Dozen, "Global Pesticide Campaigner," June 1991, p. 14.
8. "Tell EPA: It's Time to Ban Aldicarb," in PAN Update Service, September 1991.
9. Mailing of Monica Moore, PAN North America Regional Center, 30 September 1991.
10. Rae Tyson, "Target of Toxins?," *USA Today,* 24 October 1991.
11. Carolyn Pesce, "Leaks Frighten W.Va. town," in *USA Today,* 19 February 1990. See also "W.Va. chemical plant has second toxic leak in 2 weeks," *The Pittsburgh Press,* 16 February 1990.
12. From advertisement in *Chemicalweek,* 17 June 1992.
13. Quoted in Pesce, op cit.
14. "Lindane – A Global Pesticide Problem," Briefing Document for Greenpeace Press Conference, Zaragoza, Spain, 25 March 1988, p. 3.
15. "6 tons of pesticide lost in English Channel," *Philadelphia Inquirer,* 21 March 1989.
16. Demise of the Dirty Dozen, op cit.
17. Greenpeace, Dangerous Pesticide Briefing, September 1991.
18. Quoted in Rhone-Poulenc Annual Report 1990, p. 43.
19. From video "Cubatao – Valley of Death," produced by Scandinature Films, written and directed by Bo Landin, 1987.
20. "Rhodia au rythme Bresilien," in Rhone-Poulenc's internal journal *Presence,* no. 48, pp. 6 & 8.
21. Jim Brooke, "Industrial Pollution Scar's Brazil's 'Valley of Death'," *The Washington Post,* 1981.

22. Lynda Schuster, "Industrialization of Brazilian Village Brings Jobs at Cost of Heavy Pollution and Even Death," *The Wall Street Journal*, 15 April 1985.

23. Bill Hinchberger, "World's Most Polluted City Suffers Again," ENS, 2 October 1991.

24. Brooke, op cit.

25. Brooke, op cit.

26. John Frook, "The Most Polluted Place on Earth," in *Parade*, 13 December 1981.

27. "Rhodia au rythme Bresilien," op cit, p. 8.

28. Brazilian Judge Orders Closing of Rhone-Poulenc Plant," *Greenpeace Toxic Trade Update*, 6.3, 3rd quarter 1993, p. 33. Also Markus Heissler,"Welcome to a cleaner world?", in *Biotechnology and Development Monitor*, no. 18, March 1994, p. 22.

29. Heissler, ibid.

30. "Rhodia au rythme Bresilien," op cit, p. 6.

31. Ibid, p. 6.

32. From video "Cubatao – Valley of Death," op cit.

33. Ibid.

34. Quoted in Schuster, op cit.

35. Hill & Knowlton memo "Getting Results," 15 March 1988.

Source: Cartoon by Kirk Anderson.

At work among the Indians

When a boatload of fertilizer sails into Puerto Quetzal harbour, you can rest assured that Kjell Ove Lien will be standing there waiting. There are formalities to be seen to and the quality of the shipment must be checked before the next leg of the journey - transporting the fertilizer to the warehouse, then on to the coffee-growers, flower-growers and every other kind of Guatemalan farmer.

HYDRO'S MEN ON THE SPOT: Harald Johannessen and Kjell Ove Lien are key figures in Hydro's fertilizer activities in the Central American country of Guatemala.

Text: TROND AASLAND Photo: KÅRE FOSS

It seems right somehow that the man standing on the dock in the hot tropical sun is a Norwegian. That's how it has always been, and that was how it all began, with a Norwegian shipping man who went ashore there while waiting to get into the USA. At least, that's more or less how the story of Hydro's involvement in the Guatemalan fertilizer market began.

The shipping man, Torolf Johannessen, subsequently moved to Central America, became a coffee grower, started a trading company and became Hydro's agent. Johannessen is far more than just the introductory chapter in the

Source: Norsk Hydro's *Hydro Between 80 and 90* booklet regarding fertilizer, sales and use in Guatemala, December 1990.

Greenwash Snapshot #8

NORSK HYDRO a.s

A case study in synthetic fertilizer promotion.

Norsk Hydro a.s (Norsk Hydro)
Chairman: Egil Abrahamsen
Headquarters: Bygdoy alle 2, N-0240, Oslo 2, Norway
Tel: 47-2-43-21-00 Fax: 47-2-43-27-25
Major businesses: synthetic fertilizers; aluminum & magnesium; oil & gas; plastics.

Norsk Hydro is Norway's largest publicly traded company and is 51 per cent owned by the Norwegian State. Its operations span five continents. Sixty per cent of its sales are outside Norway. Norsk Hydro is a member of the WBCSD and a signer of Responsible Care and the ICC Rotterdam Charter.

In the second half of the 1980s, accidents and reports about its toxic releases catapulted Norsk Hydro into the spotlight as Norway's biggest polluter. According to a large, glossy promotional magazine, *Hydro Between 80 and 90*, such attention was "unpleasant" but "a useful lesson was learnt."[1]

If that magazine is any indication, the "lesson" Norsk Hydro learned is the benefit of greenwash. Among the dubious claims of the publication: the company simultaneously has become "a fully integrated oil company" while calling the reduction of greenhouse gas emissions "the most important environmental challenge of our time"; polyvinyl chloride is "environmentally beneficial"; and aluminum a "green metal."

Norsk Hydro devotes the most greenwash to its biggest product line: synthetic fertilizers. In a section entitled "Fertilizer Galore," Norsk

Hydro tries to demonstrate that its synthetic fertilizers serve the environment, agricultural sustainability, and food output better than organic ones.

Synthetic fertilizers are produced industrially with petroleum derived nitrogen as well as with phosphate and potash derived from geological deposits. Because they use mineral and chemical inputs, synthetic fertilizers are referred to variously as "chemical," "mineral," or "inorganic" fertilizers. With an annual capacity of 13 million tons and markets in 120 countries, Norsk Hydro is the world leader of the commercial fertilizer industry.

Synthetic Fertilizer Impacts on Air, Water, and Land

Agricultural production in general accounts for an estimated 14 per cent of global warming and the commercial fertilizer industry is an important part of this problem. Synthetic nitrogen fertilizers, which account for almost two-thirds of the world fertilizer market and 37 per cent of Norsk Hydro's fertilizer production, are of particular concern.

Nitrogen fertilizers augment the natural flux of nitrous oxide – a greenhouse gas 270 times more potent than carbon dioxide – from soil to air; fertilizing land increases nitrous oxide emissions from two to ten times the natural rate. Nitrogen fertilizers also cause the natural release of methane, another powerful greenhouse gas, to rise; synthetically fertilized rice paddies emit four times as much methane as non-fertilized ones. Overall, synthetic fertilizer manufacturing plants consume around two per cent of all commercial energy, much of which comes from carbon-emitting fossil fuels.[2]

Norsk Hydro does not address synthetic fertilizers' contribution to climate change. But it does put forth two claims regarding their impact on the water and land environment.

First, according to Gary Myers, a lobbyist for the Fertilizer Institute whom the company quotes extensively, synthetic fertilizers are the "best protector of water quality."[3] However, particularly in less-industrialized countries, such fertilizer use has led to eutrophication of rivers and waterways, making them unsuitable for marine life. Excess nitration has polluted local and urban drinking water supplies and caused a rare form of anemia in babies that the World Health Organization links to nitrogen contamination.[4]

Second, in the words of Norsk Hydro's Agricultural Division President Trygve Refvem, "it is considerably easier to develop environmentally compatible agricultural husbandry" using synthetic rather than organic fertilizer. The latter, he says, is "the least viable means of providing plant nutrients in modern agricultural practice."[5]

In fact, long-term fertility of soil – the basis of sustainable agriculture – depends on vital elements unavailable from synthetic fertilizers: organic matter that helps prevent soil erosion and loss of essential micronutrients including zinc, iron, magnesium, and boron. By replacing traditional organic fertilizers – manure, composted vegetation, plant material – synthetic fertilizers cause depletion in soil organic content and deterioration of soil structure.

Synthetic fertilizers are also widely used on monoculture plantation crops, often the so-called "high yield varieties" grown for the export market. Monoculture plantations can cause severe erosion and are extremely vulnerable to pest infestation, thus requiring heavy pesticide application.[6]

Declining soil fertility is evident, especially in the less-industrialized world. In the Punjab region of India and Pakistan, intense synthetic fertilizer use and wheat and rice monocropping have left 80 per cent of the area's soil deficient. Brazil's coffee plantations have depleted much of the country's soil past the point at which it is viable for crop production. While farmers are now experiencing diminishing returns on synthetic fertilizer use worldwide, the problem is acute in tropical countries such as Indonesia where soil organic content is relatively low.[7]

"[I] would not hesitate to say that mineral fertilizer is preferable from an environmental perspective."
 – *Gary Myers, Fertilizer Institute, in* **Hydro Between 80 and 90**[8]

"[C]hemicals can never be a substitute for the organic production of nutrients."
 – *Vandana Shiva,* **Staying Alive** [9]

To cut costs and move closer to North American and Eastern European markets, Norsk Hydro has closed fertilizer manufacturing facilities in Western Europe and acquired plants in Trinidad and Tobago and Russia. Those countries would do well to look at all sides of the fertilizer debate.

Fertilizers and Food

"A strong and reliable partner in the Third World."
– *Norsk Hydro Agricultural Division ad, 1990*[10]

Norsk Hydro predicts that consumption of its synthetic fertilizers will rise throughout the less-industrialized world. Supplying this demand is crucial to the model of agricultural production the company promotes: one which uses external inputs to maximize single crop yields destined for the export market. As the company puts it, "The African challenge consists of replacing a system of subsistence agriculture, where the farmer grows enough to feed his own family, with a system of market agriculture."[11]

This is the familiar export model banks and development agencies have advanced, in the name of ending world hunger, since industries promoted a "Green Revolution" three decades ago. Not only has this model polluted the environment and failed to maintain the nutrient cycle, it has undermined local people's self-sufficiency and reduced their access to food. The results have been devastating. According to the Earth Resources Institute, in 1970 Africa was producing enough food to feed itself; by 1984, 26 per cent of the continent's population had to eat grain from abroad, the consequence of export-oriented cash crop production in numerous African nations.[12] Today, Africa is in the midst of a continent-wide food crisis, with two-thirds of its countries experiencing chronic shortages and malnutrition. In Eastern and Southern Africa millions are starving.[13]

The costs of external inputs such as synthetic fertilizer are beyond the means of poor farmers, many of whom, unable to compete with large growers, have fallen into debt and lost their farms.[14] Less-industrialized countries that subsidize these inputs are vulnerable to cyclical price fluctuations; in Nigeria, 85 per cent of subsidies on fertilizer prices have in the past exhausted three-quarters of the agricultural budget.[15]

Africans have not been the "Green Revolution's" only victims. In India, fertilizer imports alone have in some years eaten up 20 per cent of the country's export earnings.[16]

Large areas in South Asia have suffered tremendously; in 1987 more children died from malnutrition in India and Pakistan than in all of Africa.[17] And in Latin America, food consumption per person declined from 1980 to 1990, while infant mortality – an indicator of nutritional deficiency – is rising in many countries.[18] Despite an increase in fertilizer consumption of two and a half times from the early 1970s to the late 1980s, imported food aid in cereals leapt more than four-fold in Latin America and the Caribbean.[19]

Norsk in Guatemala

As the ad reproduced on page 118 shows, Norsk Hydro uses Guatemala to exemplify the geographic reach of its synthetic fertilizer distribution, exhibiting a photograph of a highlands Indian farmer and daughter transporting produce on a bicycle. Norsk Hydro established a foothold in Guatemala with the help of a coffee growing and exporting companies and now uses 100 dealers to control ten to 15 per cent of Guatemala's fertilizer market. Coffee, the country's largest export crop, has always accounted for most of the sales, but Norsk Hydro is moving into Guatemala's highlands to sell fertilizer for crops such as sugar and cotton, also heavily exported.[20]

The growth of coffee, cotton and sugar production has dispossessed Guatemala's Indians of their most fertile land, with large plantations displacing the Indians' small subsistence plots. These Indians are among the hemisphere's poorest people, with a per capita income that is seven per cent of Guatemala's already low average.[21] Infant mortality in the Indian areas of the highlands is 160 per 1000, a rate double that of the country's general population.[22]

Norsk Hydro's Guatemalan operations are an example of the kind of foreign investment which has discouraged production of basic foodstuffs for domestic consumption. While fertilizer use doubled from the early

1970s to 1988, imported cereals food aid rose over 35 times.[23]

The country depends on agro-exports which generate 66 per cent of export income and account for 25 per cent of its Gross Domestic Product. However, decreases in real export value caused the GDP to drop from the early to mid-1980s, while external debt rose. Guatemala is currently in a major fiscal crisis resulting largely from its extreme dependence on cash crops.

These macroeconomic indicators mean great human misery. According to UNICEF, Guatemala is Central America's most impoverished country and has the lowest "physical quality of life index" in the region. An estimated 60 per cent of the population lives in absolute poverty and one national survey showed that only 27 per cent of all children between six months and five years had normal physical development. The standards of rural subsistence are arguably lower today than they were in the colonial period 350 years ago.

> "Seen from my standpoint, there are many lamentable things here, but I realize that I must bear in mind that I'm here solely to promote the sale of Hydro's products."
> – *Kjell Ove Lien, Norsk Hydro's
> Guatemalan representative (pictured on page 118), 1990*[24]

Norsk Hydro and Energy

Norsk Hydro was one of 13 large industrial companies that in the late 1980s signed secret contracts with Hydro-Quebec, builder of the huge James Bay hydroelectric project which has threatened to destroy wilderness and displace 18,000 indigenous people from an area the size of France. Hydro-Quebec agreed to sell electricity to the companies at prices well below production cost in order to boost energy demand artificially to justify construction of the project's second phase, which was cancelled in 1994.[25]

The deal with Hydro-Quebec, to provide electricity to Norsk Hydro's US$550-million magnesium plant in Becancour, Canada (its single biggest industrial project outside Norway), became public in 1991. This contract effectively subsidized Norsk Hydro's investment, saving the company US$25-30 million in energy fees compared to those paid by other customers in Quebec.[26] Analysts estimated that Hydro-Quebec's initial losses from the secret contracts could amount to as much as US$220 million per year, a cost which its other customers would incur.[27]

The US Department of Commerce argued that the deal between Hydro-Quebec and Norsk Hydro was an unfair subsidy and threatened to impose a large duty on magnesium exports of Norsk Hydro's Becancour plant. In 1992, under this pressure, the two companies amended their contract. Norsk Hydro would no longer receive annual rebates of up to 50 per cent of its electricity bills, but it would be able to purchase power at rates pegged to the world price of magnesium.[28] In October 1992, the Commerce Department made a preliminary ruling to reduce the duty significantly.

Norsk Hydro is a major producer of aluminum and aggressively markets the metal for its environmental benefits (see Alcoa case study for more information about the environmental problems of aluminum).

Endnotes

1. *Hydro Between 80 and 90,* Profile Magazine, published by Norsk Hydro, December 1990, p. 62.
2. Dr. Arjun Makhijani, *Climate Change and Transnational Corporations Analysis and Trends*, United Nations Centre on Transnational Corporations, New York, 1992, p. 91. See also Edward Goldsmith and N. Hildyard, "World Agriculture: Toward 2000 FAO's Plan to Feed the World," in *The Ecologist,* vol. 21, no. 2, March/April 1991, pp. 89-90.
3. *Hydro Between 80 and 90,* p. 55.
4. *Climate Change,* pp. 94 & 96.
5. *Hydro Between 80 and 90,* p. 45.
6. For information about the problems of synthetic fertilizer use see: Goldsmith and Hildyard, op cit, esp. pp. 85-87 & 92; Vandana Shiva, "The Green Revolution in the Punjab," in *The Ecologist,* vol. 21, no. 2, March/April 1991; and Vandana Shiva, *Staying Alive: Women, Ecology, and Development,* Zed Books, London, 1989, esp. pp. 138-152.
7. For information in this paragraph see: John Madeley and E. Hawksley, "World Food Production," in *South Magazine,* January 1990, p. 43; Goldsmith and Hildyard, op cit, pp. 86-87.

8. *Hydro Between 80 and 90,* p. 55
9. Shiva, op cit, p. 145.
10. The ad can be found in *South Magazine,* January 1990.
11. *Hydro Between 80 and 90,* p. 54.
12. Shiva, op cit, p. 138.
13. Editorial from March/April 1991 edition of *The Ecologist,* p. 43.
14. *Climate Change,* p. 94.
15. Goldsmith and Hildyard, op cit, p. 91.
16. Madeley and Hawksley, op cit.
17. *Ecologist* editorial, p. 43.
18. Lester Brown, "The Illusion of Progress," in *State of the World 1990,* a Worldwatch Institute Report, Washington, DC, 1990, p. 4.
19. See figures from*Poverty: World Development Report 1990*, published for The World Bank, Oxford University Press, pp. 184-185; and Brown, ibid.
20. *Hydro Between 80 and 90,* pp. 40-41.
21. Walter LaFeber, *Inevitable Revolutions*, Norton, New York, 1983, esp. pp. 8-9.
22. Unless otherwise referenced, for this and the following information on the situation in Guatemala see Marcus Colchester, "Guatemala: The Clamour for Land and the Fate of the Forests," in *The Ecologist*, vol. 21, no. 4, July/August 1991, esp. pp. 177-182.
23. *Poverty*, op cit, p. 184.
24. *Hydro Between 80 and 90*, p. 41.
25. Graeme Hamilton, "Hydro to release summary of secret contracts: Drouin," and "Norsk subsidized by cheap power," both in *The Gazette (Montreal)*, 12 & 30 April 1991.
26. Bernard St. Laurent, "Bourassa on wrong side of battle over Hydro contracts," *The Gazette*, 22 April 1991.
27. Graeme Hamilton, "Contracts could cost Hydro up to $220 million: analyst," *The Gazette*, 8 May 1991.
28. Leo Ryan, "Quebec Lauds US Cut of Norsk Hydro Duty," *The Journal of Commerce,* 15 October 1992.

Citizen's Guide to Greenspeak

The citizen activist should have a good understanding of the language of greenwash: GREENSPEAK. Here are some helpful definitions:

Excellent environmental performance: Having fewer accidents while manufacturing deadly products.

Product Stewardship: Asking sales personnel to remind poor agricultural workers to use expensive, heavy protective gear in 40°C heat.

Recycling: Taking the waste from a dangerous process and turning it into a useless product.

Continuous Improvement: Industry has a terrible environmental record, but that will always be due to past practices.

Corporate environmentalism: Environmental laws and citizen watchdogs are no longer necessary.

Change in corporate culture: Prettier advertisements.

Business Council For Sustainable Development: Council For Sustainable Business Development.

Source: *Chemicalweek*, 17 July 1991.

Greenwash Snapshot #9

*SANDOZ LTD**

A case study in dirty industry transfer and biotechnology experimentation.

Sandoz Ltd. (Sandoz)
Chairman & Chief Executive Officer: Marc Moret
Headquarters: CH-4002, Lichstrasse 35, Basel, Switzerland
Tel: 41-61-324-11-11   Fax: 41-61-324-80-01
Major businesses: pharmaceuticals; chemicals; agrochemicals; seeds; nutrition.

Sandoz has operations in 55 countries and is a signer of Responsible Care and the ICC Rotterdam Charter. It is a member of the WBCSD.

"[T]he Rhine is now dead. The whole ecosystem is destroyed due to this accident."
– *Walter Herrmann, Chief Inspector, Rhine River Police, 1986* [1]

"We didn't think about the Rhine."
– *Ernst Wessendorf, Sandoz information officer, 1986* [2]

The choice of a pristine river for Sandoz's image is conspicuously ironic. Sandoz is well known as the company responsible for the worst river spill in history.

During a 1986 fire at Sandoz's production facility near Basel, Switzerland, up to 30 tons of extremely hazardous organophosphate

---

* *In early 1996, Sandoz and Ciba agreed to merge, forming a new company, Novartis.*

pesticides called disulfoton and parathion washed into the Rhine River. The poisons killed fish, wildlife, and plants for hundreds of miles. Known locally as "Chernobale" (Bale is the French name for Basel), the Rhine spill was the most notorious corporate industrial accident in Western Europe since F. Hoffmann-La Roche & Co.'s Seveso plant released dioxins over the Italian countryside in 1976.[3]

Following the spill, Sandoz "cleaned up" its operations by moving 60 per cent of its organophosphate production to Resende, Brazil. In 1989, shortly after another ton of Sandoz disulfoton nearly spilled into the Rhine, Sandoz moved the rest of its organophosphate production to Brazil.[4] Apparently, the company believes that it is legitimate to move a production process which severely harmed a river ecosystem in Europe to Latin America.

Sandoz "Tests" Biotechnology Abroad

Sandoz is one of many TNCs involved in biotechnology, and genetic engineering, which in theory allows genes from one organism to be moved into another. Biotechnology is used in a variety of industries, including agriculture, chemicals, and pharmaceuticals.[5] Sandoz's Austrian subsidiary Biochemie GmbH is now producing recombinant Bovine Growth Hormone (rBGH, also referred to as Bovine Somatotropin or BST) under a license from Monsanto and exporting it, probably to Central and Eastern Europe. rBGH is designed to increase milk production in cows.

rBGH is the first genetically engineered agricultural product to reach the market. It is not licensed in Switzerland or Austria and in December 1994 the European Union (EU) extended the moratorium on the product until the beginning of the next century.[6] The reason behind this caution is simple; rBGH has not been approved for use in many countries because its risks have not been fully investigated.

The proposal to extend the moratorium on rBGH, however, has come under fire since the completion of the Uruguay Round of the General Agreement on Tariffs and Trade (GATT) in December 1993. GATT allows a country to challenge another country's domestic law if that law is seen as a barrier to trade; a challenge to the EU moratorium on rBGH is probable.[7]

Opposition to rBGH

High-producing milk cows have shown elevated susceptibility to infectious diseases. If rBGH use increases health problems in cattle, antibiotic use will also rise. Cows can accumulate antibiotic resistant organisms which, when transferred to humans through milk, cause serious infections which are difficult to treat precisely because of antibiotic resistance. rBGH's effects on infants are of particularly great concern.[8] Because of its potential hazards, Denmark, Sweden, and Norway have banned rBGH. Three provinces in Canada have prohibited the sale of rBGH milk and a national "Pure Milk Campaign" has been launched to block the licensing of rBGH.

A national coalition of farmers and consumers in the United States is organizing against companies which market rBGH.[9] In November 1993, the US Food and Drug Administration approved rBGH use in milk cows, a move which consumer and farm advocacy groups have strongly criticized and are resisting. The struggle is being fought through the legislatures in some states to label products which have been derived from cows treated with rBGH or which are rBGH free appropriately (for more information on rBGH see Monsanto case study).[10]

While public opposition to rBGH grows in the industrialized North, biotechnology companies are using Central and Eastern Europe as well as countries in Africa, Latin America, and Asia as testing grounds for genetically engineered products whose safety is unproven. In the case of rBGH, the financial rewards go to Sandoz and other TNCs, and the health risks to the farmers, consumers,and children of less-industrialized nations.

The "Gene Revolution": Cui Bono?

Biotechnology companies promote ventures in genetic engineering like rBGH as a solution to malnutrition.[11] Such claims have disturbing similarities to those made several decades ago. In the 1960s, corporations promised that a new generation of pesticides and other agricultural inputs would help the poor. But the "Green Revolution" failed to eradicate

hunger and the heavy use of toxic chemicals has contributed to severe,
long-lasting environmental and human health problems. The "Green
Revolution" did succeed, however, in making profits for agrochemical
companies.

The "gene revolution" may be more of the same. rBGH and other
genetically engineered products are manufactured and patented to be sold
at high prices, often back to the same countries where genetic stock
originated. While the environmental and social benefit of these products
is dubious, their profit-making potential for TNCs is considerable.

There are many reasons to question the promise that new
biotechnologies will feed the hungry. The case of rBGH, which may well
be incompatible with the needs of those it is supposed to help, is
an example:

- rBGH is expensive and requires sophisticated accessories such as
  refrigerated stables. Because of its high cost, rBGH will be impossible to use for many small-scale farmers who are most in need of the
  income and nutrition that milk production provides. Only wealthy
  cattle owners will be able to afford rBGH, and as larger, more modern
  farms increase production small and medium farmers will be forced
  out of business.[12] Further, and also because of its cost, rBGH will
  probably reach less-industrialized countries through development-
  aid packages which are tied to buying goods and services from
  companies in the donor country.

- rBGH's use may limit milk's accessibility. The concentration of
  cattle on large farms may benefit some urban dwellers. In rural
  regions, however, a wide distribution of cows among the population
  is needed to increase access to cattle products such as milk.

- Greater milk production can displace traditional, cheaper food and
  protein sources. In many countries, concentration of cattle can and
  does lead to the conversion of agricultural land to pasture and the
  feeding of grain to cattle rather than to hungry people.[13] It should be
  noted that a significant portion of the world's adult population is
  unable to drink milk because of lactose intolerance.

- rBGH use will harm the environment. The urine and manure generated on large cattle feedlots can pollute surface and groundwater as well as soil which could be used for agriculture.[14]

rBGH is typical of biotechnologies in that its development and marketing are controlled in the private sector, by transnational corporations such as Sandoz. "The issue of privatisation is increasingly becoming a threat to democracy and people's will," Vandana Shiva has written, "as the same scientists work on contract for TNCs, function on government regulatory bodies, and dominate scientific research. In this context it is up to citizens, free of TNC and government control, to keep public issues and priorities alive, and have a space for public control of the new biotechnologies."[15]

Endnotes

1. Quoted in Thomas Netter, "Tide of Protest Rises on Rhine," *International Herald Tribune*, 12 November 1986.
2. Quoted in Margaret Studer, "Rhine Spill Punctures Swiss Complacency About the Advanced State of Their Society," *The Wall Street Journal,* 13 November 1986.
3. Studer, op cit.
4. From a Greenpeace communication with a Sandoz employee. See also "Die Katastrophe als Gluckfall," in *Weltwoche*, 31 October 1991. And Ernest Beck, "After the Party was Over, Poisoned Eels on a Silver Plate," Gemini News Service, 25 November 1986.
5. Vandana Shiva, *Biotechnology and the Environment,* Third World Network, Malaysia, pp. 1-2.
6. Gen-und Biotechnologie, Nutzungsmoglichkeiten und Gefahren-potentiale Handlungsbedarf fur Osterreich zum Schutz von Mensch und Umvelt, Bundesministerium fur Umvelt, Jugend und Famine, p. 255. "Bio/Technology/Diversity Week," vol. 3, no. 7, 25 March 1994, produced by the Institute for Agriculture and Trade Policy. "EU extends rBGH Ban," *Multinational Monitor,* January/February 1995.
7. "Bio/Technology/Diversity Week," ibid.
8. For more information of the potential hazards of rBGH use see: Nandini Katre, "A Case Against BST/rBGH," Greenpeace International, 1992;

geneWatch, v. 7, nos. 1-2, pp. 3-5; Shanti George, "Is Biotechnology the Solution and to What Problems?: Bovine Somatropin," for the Berne Declaration Group: World Food Day, October 1991; and "Third World Marketing & Promotion of Biosynthetic Hormone," Rural Advancement Fund International Communique, October 1990.

9.   Shiva, op cit, p. 8.

10. "Bio/Technology/Diversity Week," op cit, and Keith Schneider, "Maine and Vermont Restrict Dairies' Use of a Growth Hormone," in *The New York Times,* 15 April 1994.

11. *geneWatch,* p. 4, and Karol Cutler, "Getting Biotechnology to Third World Nations: Monsanto's New Program," in *AG Biotechnology News*, November/December 1990, p. 10.

12. Katre, *geneWatch*, p. 3, and Shiva, p. 8, op cit.

13. Katre, *geneWatch*, p. 4, op cit.

14. *geneWatch,* op cit, p. 3.

15. Shiva, op cit, p. 26.

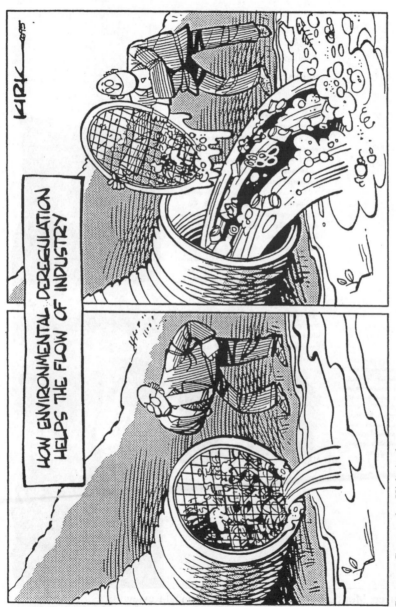

Source: Cartoon by Kirk Anderson.

GROUNDWATER PROTECTION

IT IS OUR PLEDGE TO ENSURE GROUNDWATER SAFETY.

Our approach starts in the plants, where we engineer new facilities for groundwater safety. Groundwater assessments are conducted routinely for all major plant sites worldwide and other plants with known groundwater concerns. Each of our plants maintains a groundwater protection plan, including inspection, testing and maintenance of facilities to prevent groundwater contamination.

Monsanto's commitment to groundwater protection extends beyond plant boundaries. We recently completed the largest rural well-water survey of its kind ever conducted in the United States. We were looking for traces of commonly used herbicides in drinking water.

The study covered more than 1,400 wells in 26 states. The results: More than 99 percent of the wells show no contamination from Monsanto herbicides.

Demonstrating our commitment to groundwater safety, we offer analytical and financial assistance to correct any rural well in the United States where our herbicides are found in excess of U.S. Environmental Protection Agency (EPA) standards.

Doug Rushing, a field environmental support manager, meets with a customer in rural Iowa to discuss Monsanto's rural well assistance program, which provides analytical and financial support for evaluating the quality of well water.

10

Source: Monsanto 1991 Environmental Annual Review.

Greenwash Snapshot #10

MONSANTO CORPORATION

A case study in greenwash science.

Monsanto Corporation (Monsanto)
Chairman & Chief Executive Officer: Robert B. Shapiro
Headquarters: 800 N. Lindbergh Boulevard St. Louis, Missouri 63167, USA
Tel: 314-694-1000 Fax: 314-694-7625
Major businesses: chemicals; pesticides; pharmaceuticals; biotechnology.

With facilities in at least 12 countries, Monsanto earns 37 per cent of its revenue outside the US. Monsanto is a member of Responsible Care.

In a 1989 speech which introduces some of the tortured logic behind corporate "self-regulation," former Monsanto Chairman & Chief Executive Officer Richard Mahoney argued that numerous environmental laws and regulations were ineffective in improving environmental performance and emphasized "voluntary efforts which might achieve more than Congress could imagine." For Mahoney, the rationale behind such efforts is "public perception." "Rightly or wrongly," he said, "the public perceives that environmental problems are growing worse. To earn the right to operate, my company must respond to the public's perception, and the public's concern."[1]

Mahoney recognizes the existence of public concern, but without acknowledging the validity of its source. This is a nearly explicit admission that he sees public relations improvements as more important than actual environmental improvements. (This attitude – really an ambivalence about whether environmental problems are real or perceptual – is also prevalent in the industry-wide Responsible Care Program.)

Monsanto's environmental programs and documents reflect this prioritization of imagined over substantive environmentalism. Most conspicuous is the technique of scientific greenwash, that is, using faulty science to justify or cover up harmful activities.

Science for Sale

1. *Dioxin*

> "For better or worse, we're deeper into 'dioxin[s]' than anyone, even though the public links them more to Dow."
> –   *Monsanto internal memo made public in 1987* [2]

As the third largest US chemical company and the inventor of PCBs, Monsanto manufactures, uses, and disposes of vast quantities of chemicals associated with the generation and dispersal of dioxins. Dioxins are the among the most notorious toxins ever made, and are present in the general population and widespread in the environment.

Threatened with billions of dollars of damages and business losses, dioxin producers have crusaded to convince government officials and the public of dioxin's safety, and Monsanto has led the charge. About dioxin-related court cases in the 1980s, a Monsanto memo asserted, "If we win them all, there won't be others. If we lose them all, Katy bar the door."[3]

From 1980 to 1984, dioxin studies conducted by Monsanto researchers or academics funded by Monsanto appeared in scientific journals including the *Journal of the American Medical Association*. Based on research of workers exposed to dioxin at Monsanto's herbicide plant in Nitro, West Virginia, the studies found no significant negative health effects other than a skin condition called chloracne.

Released to the press with great fanfare and touted as the most comprehensive investigations into dioxin's health effects, the studies played a major role in decisions regarding regulation, liability, and victims' compensation by the US Environmental Protection Agency (EPA), the US Veteran's Administration, and in Europe, Australia, and New Zealand. They are the basis for frequently-heard claims that not one single death can be attributed to dioxin.[4]

A few critics expressed initial skepticism at Monsanto's "science."[5] But in litigation ending in 1989 (Kemner v. Monsanto), evidence emerged which suggested a pattern of manipulated data and misleading design in Monsanto's Nitro studies.

According to testimony from the trial, Monsanto misclassified exposed and non-exposed workers, arbitrarily deleted several key cancer cases, failed to verify classification of chloracne subjects by common industrial dermatitis criteria, did not provide assurance of untampered records delivered and used by consultants, and made false statements about dioxin contamination in Monsanto products.[6]

"There are numerous...flaws in the Monsanto health studies. Each of these misrepresentations and falsifications served to negate any conclusions of adverse health effects from dioxins."
– Dr. Cate Jenkins, US EPA Regulatory
   Development Branch, 1990[7]

"There is a clear pattern of fraudulent misconduct in the dioxin science performed by the chemical industry and its indentured academics."
– Dr. Samuel Epstein, Professor of Occupational and
   Environmental Medicine, University of Illinois, 1990[8]

In 1990, the Australian federal government announced a review of decisions based on the Nitro studies' results. That same year, the EPA's Office of Criminal Enforcement made preliminary moves to investigate Monsanto's conduct in the Nitro studies and other dioxin-related issues.[9] This investigation continued for two years before being quietly closed in 1992 "without," according to a July 1994 internal EPA analysis of the process, "ever determining or even attempting to determine if the Monsanto studies were valid or invalid, let alone fraudulent. Instead, it investigated and illegally harassed the whistleblower, Cate Jenkins."[10]

Because the investigation had failed to answer the question about whether or not Monsanto had manipulated the studies to reduce liability to the Vietnam veterans, the internal analysis recommended, the EPA should "convene a panel of disinterested scientists, with the full support and authority of the US government, to determine if the studies are valid, and if not then determine whether the results would have been positive if the studies were done correctly."[11]

A 1991 study by the National Institute of Occupation Health and Safety (NIOSH) revised Monsanto's "scientific" methodology and reversed its conclusions. Examining the causes of death in workers at the Nitro plant and other chemical facilities, NIOSH found a statistically significant increase in cancers in the workers at all sites when dioxin-exposed workers at Monsanto's plant and elsewhere were examined as an aggregate group.[12] That is, NIOSH indicated that dioxin exposure **did** increase the likelihood of cancer.

Moreover, 1992-94 documents by the US EPA suggest that the weight of evidence indicates that dioxin can be considered a human carcinogen. EPA researchers have estimated that dioxin exposure currently poses cancer hazards which are 100 to 1000 times greater than the standard "acceptable" risk of one cancer per million. According to such risk estimates, dioxin would cause 350-3500 cancers annually in the US, or up to three per cent of all cancers.[13]

2. *Pesticides*

Monsanto funds greenwash science for hazards other than dioxins. The company is a sponsor of the Council for Agricultural Science and Technology (CAST), a group which Ciba, DuPont, Dow, among others, also support, and whose purported mission is to provide "unbiased, scientific information concerning food and agriculture."[14] Monsanto's motivation in funding this research stems from its position as a leading pesticide maker in the US, and the controversial nature of some of its products including butachlor, alachlor, and glyphosate (see "Additional Information" below).

Trying to deflect attention from these kinds of issues, CAST's "impartial" research maintains that pesticide residues in food pose no human health risk and alternatives to pesticide use are infeasible. These conclusions belie CAST's claim to objectivity and reveal it to be a pro-industry advocacy group which employs "science" to greenwash the harmful activities of members like Monsanto.

"I was told by a scientist who was involved with CAST that CAST convened at least once and maybe more than that to create responses to announced television programs on pesticide residue or other problems. They began the meetings with the clear notion that they were going to repudiate whatever was said on the television program....It was, 'Prepare a response to discredit the television show.' That doesn't sound too objective to me."
— *James Turner, consumer attorney, 1991*[15]

"They [CAST] continue to turn a blind eye toward scientific evidence. They are certainly not an unbiased group; rather they are apologists for the industry."
— *Michael Colby, Food & Water, Inc., 1991* [16]

3. *Biotechnology*

Monsanto is one of four TNCs developing recombinant Bovine Growth Hormone (rBGH). The company has manufacturing facilities in the US, Austria, and the former Soviet Union. Because of many questions about rBGH's health risks, for people as well as cows, the US only approved the hormone in late 1993 and the European Union has a moratorium on rBGH through the beginning of the next century. This absence of regulatory approval in the industrialized world did not, however, prevent Monsanto from obtaining approval for rBGH in less-industrialized countries where use of the product may be particularly inappropriate and hazardous (see Sandoz case study for more on the problems with rBGH).[17]

Nor did lack of regulatory approval stop Monsanto from promoting rBGH as safe and effective. In 1991, the US Food and Drug Administration (FDA) said the company's materials on rBGH "go beyond the legitimate exchange of scientific information" and ordered Monsanto to stop making unsubstantiated claims about rBGH.[18]

In November 1993, despite strong opposition from US consumer and farm advocacy groups, the FDA finally approved rBGH for use on cows in the US. But as of late 1994, only seven per cent of the country's dairy farmers were using rBGH.[19] Furthermore, several big supermarket chains and dairy cooperatives have refused to purchase dairy products from rBGH-treated cows, and there are reports from three US states that cows injected with rBGH have suffered serious udder infections and even death which some farmers say was caused by the drug.[20] Both Monsanto and the FDA are resisting efforts to label milk about rBGH.[21]

As with the dioxin studies, an economic motive has driven Monsanto's actions. Since 1984, the company has already spent over US$300 million on rBGH R&D and in its 1991 Annual Report complained that rBGH's costs were affecting "financial results adversely."[22] Monsanto's projected annual earnings from rBGH are US$300-500 million in the US and US$1 billion worldwide.[23]

Additional Information

Pesticides

Although it no longer makes Agent Orange or parathion, Monsanto markets some of the world's top-selling herbicides: glyphosate (trade name Roundup) and alachlor (trade name Lasso). Glyphosate causes kidney tumors in lab animals. Human exposure to Roundup has resulted in nausea, skin and eye inflammation, bronchial constriction, and nervous system disorders. Alachlor, which the EPA considers a carcinogen, has caused lung, stomach, and nasal tumors in lab animals and has contaminated over 46,000 US drinking water wells.[24]

Monsanto also produces butachlor (trade names Machete, Lambast),a herbicide which poses both acute and chronic health risks and can contaminate water supplies. Although Monsanto manufactures butachlor in Iowa, the herbicide has never been registered in the US or gained a food residue tolerance. In 1984, the EPA rejected Monsanto's registration applications due to "environmental, residue, fish and wildlife, and toxicological concerns." Monsanto refused to submit additional data requested by the EPA.[25] Despite its recognized dangers, Monsanto sells butachlor abroad. Dozens of countries in Latin America, Asia, and Africa use the product, primarily on the paddy rice which constitutes almost all of US rice imports.

Accidents and Hazardous Waste

Monsanto invented polychlorinated biphenyls (PCBs) and for many years was the world's sole PCB manufacturer. PCBs are a group of chemicals so toxic that their production has been banned virtually worldwide but they continue to threaten the health of marine mammals, birds, humans, and even entire ecosystems.

According to EPA data, Monsanto consistently ranks as one of the largest corporate generators of toxic emissions into the US environment. Throughout the country, this has had a number of negative human health and ecological consequences. Some examples:

• 1986 – A US District Court found Monsanto liable in the death of a Texas employee from leukemia allegedly caused by exposure to the carcinogenic benzene. The plaintiff's family contended that Monsanto had neglected to monitor benzene emissions at the plant and had failed to instruct workers about the risks of handling benzene-tainted compounds. The court awarded the plaintiff's family US$108 million.[26]

• 1988 – Monsanto agreed to a US$1.5 million settlement in a chemical poisoning case filed by over 170 former employees of the company's Nitro, West Virginia facility. Six workers said they had been exposed to chemicals which allegedly gave them a rare form of bladder cancer.[27]

• 1990 – Monsanto paid US$648,000 to settle charges that it allegedly failed to report significant risk findings from health studies to the EPA as required under the Toxic Substance Control Act.[28]

• 1991 – The Massachusetts Attorney General's office fined Monsanto US$1 million – the largest ever assessed in Massachusetts for violation of a state environmental law – for illegally discharging 200,000 gallons of acid-laden wastewater from a plant and failing to report the release immediately as well as understating the volume of the release. According to the state's Department of Environmental Protection, Monsanto, which paid a US$35,000 fine in 1988 for failing to report an acid spill at the same facility, had a history of violating spill-reporting laws.[29]

• 1992 – Monsanto agreed to pay US$39 million of a US$208 million Superfund settlement with 1700 Houston residents who claimed injuries as a result of living near a former toxic waste dump, labeled one of the worst such sites in Texas. Plaintiffs argued that Monsanto deposited 519 million pounds of hazardous compounds into unlined holes in the ground. Children in the area suffer health problems including immune deficiency disorders, cancer, and facial deformities allegedly due to exposure to toxins leaking from the site.[30]

> "We pledge to ensure that no operation poses any undue risk to employees, the community, or the environment....Monsanto will keep the community and employees fully informed of any significant hazards."
> – from the "Monsanto Pledge", 1990 [31]

The Mississippi River has suffered especially from the company's pollution. Monsanto's Sauget, Illinois plant has discharged an estimated 34 million pounds of toxins annually into the river. The facility is a major producer of chloronitrobenzenes, bioaccumulative teratogens detected at levels as high as 1000 parts per billion in fish over 100 miles downstream. Before they were banned in the 1970s, the Sauget plant was the world's only manufacturer of PCBs. Besides being present at high levels in Mississippi fish, PCBs are ubiquitous in the global ecosystem.[32]

Monsanto's Muscatine, Iowa plant, which produces alachlor, butachlor, and other highly toxic compounds, releases at least 265,000 pounds of chemicals per year directly into the Mississippi. According to the US Fish and Wildlife Service: "[T]he combined effect of the Monsanto discharge with other discharges may severely stress and degrade the [aquatic] habitat."[33] Agricultural chemicals in the discharge were of particular concern.

Endnotes

1. From "Breaking the Circle of the Short-Term Fix–We Must Earn the Right to Operate," remarks by Monsanto CEO Richard Mahoney at EPA Region III Environmental Conference, 16 November 1989.
2. "Marathon Trial Spill Nears End in Illinois After 3 1/2 Years," *The New York Times,* 19 August 1987.
3. Ibid.
4. Joe Thornton, Greenpeace, *Science for Sale: Critique of Monsanto Studies on Worker Health Due to Exposure to 2,3,7,8-Tetrachlorodibenzo-P-Dioxin(TCDD)*, November 1990, p. 3.
5. A. Hay & E. Silbergeld, letter, "Assessing the risk of dioxin exposure," Nature, 315: 102-103, 1985.
6. Thornton, op cit, esp. pp. 3-4; Dr. Cate Jenkins, EPA, "Criminal Investigation of Monsanto Corporation – Cover-up of Dioxin Contamination in Products – Falsification of Dioxin Health Studies," November 1990; Affidavit of Cate Jenkins, Ph.D., "Recent Scientific Evidence Developed after 1984 Supporting a Causal Relationship Between Dioxin and Human Health Effects."
7. Jenkins, "Criminal Investigation," op cit, p. 5.
8. Press Release "Greenpeace Exposes Fraud in Industry Dioxin Studies; Calls for EPA Actions to Prevent Dioxin Pollution," 29 November 1990.
9. "Science for Sale," op cit, p. 10.
10. William Sanjour, Policy Analyst, US EPA, Memorandum "The Monsanto Investigation" to D. Bussard, 20 July 1994, p. 2.
11. Ibid, pp. 23-24.
12. Jenkins, "Criminal Investigation," op cit, p. 5.
13. See Joe Thornton, Greenpeace, *Achieving Zero Dioxin – An Emergency Strategy for Dioxin Elimination,* Washington, DC, July 1994, pp. 15-16.
14. Mark Megalli & Andy Friedman, "Masks of Deception Corporate Front Groups in America," in *Essential Information,* December 1991.
15. Ibid.
16. Ibid.
17. Nandini Katre, Greenpeace International, "A case against BST/BGH," January 1992.
18. "Monsanto Told to Stop Promoting Milk Hormone," *Business Information Wire*, 13 February 1991.
19. Keith Schneider, "Despite Critics, Dairy Farmers Increase Use of a Growth Hormone in Cows," in *The New York Times,* 20 October 1994.

20. Ibid.
21. Peter Montague, *Rachel's Hazardous Waste News,* issue 381.
22. Monsanto 1991 Annual Report, pp. 29 & 30.
23. Montague, op cit.
24. Esty Dinur, "Roundup-Is It Good for You and Your Environment?", *World News,* 23 December 1991; Northwest Coalition for Alternatives to Pesticides, "Roundup (Glyphosate) Information Packet," 1985; Council on Economic Priorities (CEP), Profile of Monsanto Company, draft, 1991, pp. 4-5; Corporate Profiles: Monsanto, *Multinational Monitor,* December 1990, pp. 13-14; Pesticide Action Network, "Monsanto Well Contamination Cleanup," *Global Pesticide Campaigner,* May 1992, pp. 15-16.
25. Greenpeace, *Never-Registered Pesticides: Rejected Toxics Join the 'Circle of Poison,'* February 1992, pp. 10-12.
26. CEP, op cit, p. 11.
27. CEP, op cit, p. 12.
28. CEP, op cit, p. 11.
29. "Monsanto Agrees to Pay $1 Million Penalty to Massachusetts for Acid Release," *Hazardous Materials Intelligence Report,* 18 January 1991.
30. "Companies Settle Houston Waste Site Claims with Residents for More than $200 Million," *Toxic Materials News,* 24 June 1992; Susan Fahlgren, "Superfund Settlement," AP, 24 June 1992.
31. John Dushney, Manager, Monsanto Everett, MA plant, in *The Monsanto Mystic View Plus,* April 1990, p. 2.
32. Greenpeace, *We All Live Downstream: The Mississippi River and the National Toxics Crisis,* December 1989, p. 68.
33. Ibid, p. 62.

ALL OF OUR MILK & CREAM COMES FROM FAMILY FARMERS WHO AGREE NOT TO USE rBGH.

rBGH is a synthetic growth hormone recently approved for dairy cows by the FDA

This label, formerly used by Ben & Jerry's on its ice cream cartons, was withdrawn by the company because four states prohibited such labels, and several other states threatened to sue the company.

Dear Customer
Since 1926 Sunnydale Farms has taken pride in the high quality of our products. We recognize the concern our customers have about rBST in their milk. We have obtained the agreement of our farmers and suppliers to ship raw milk to Sunnydale Farms Processing Plant from cows not treated with rBST. This applies to all our milk.
The FDA has approved the use of rBST and found no significant difference between milk from rBST treated cows and those not so treated.

SUNNYDALE FARMS

Labels like this are still used by several milk companies in New York and other states. Monsanto opposes labelling of milk with rBGH, genetically engineered soybeans, and other genetically engineered foods.

Bill Vorley heads the Farmer Support Team, helping develop training programs in developing countries for the correct use of plant protection products. Training is conducted by local company employees, for example at Ciba-Geigy's Santa Rosa Training Centre, the Philippines.

Source: Ciba 1991 Annual Report.

Greenwash Snapshot #11

CIBA LTD*

A case study in pesticide contamination and the potential hazards of biotechnology.

Ciba Ltd. (formerly Ciba-Geigy)
Chairman, Executive Committee: Heini Lippuner
Headquarters: CH-4002, Basel, Switzerland
Tel: 41-61-696-11-11 Fax: 41-61-696-43-54
Major businesses: agrochemicals; seeds; pharmaceuticals; dyestuffs; plastics; pigments.

Ciba has facilities in 60 countries and sells almost all of its products outside Switzerland. Ciba is a member of the WBCSD and a signer of the ICC Rotterdam Charter and Responsible Care.

> "All we ask...is that there should be less grudging recognition that we have contributed to solving some of the world's food problems and occasional recognition that we are indeed responsible members of society."
> –   *T.W. Parton, President*
>   *Ciba's Agricultural Division, 1988* [1]

The self-pitying tone of the executive quoted above makes one wonder why Ciba, the world's top agrochemical company, was ever perceived as irresponsible. There are reasons. In 1975, Ciba tested the safety of the insecticide monocrotophos (trade name Nuvacron) by spraying it on 40 children and adults in India. The World Health Organization has classified monocrotophos as "highly hazardous." "Such tests are unheard of,"

---

*   *In early 1996, Sandoz and Ciba agreed to merge, forming a new company, Novartis.*

an Indian commentator asserted, "...I doubt that Ciba would have done these tests in either Europe or the United States."[2]

Also contributing to Ciba's reputation for irresponsibility is a 1976 "field trial" on six Egyptian children with the pesticide chlordimeform (trade name Galecron), a carcinogen. The children later suffered poisoning symptoms. Even in 1976, Ciba's own studies on Galecron showed "toxicological effects that we consider to be grave." In Europe the company warned that children should be kept away from the product. Nonetheless, Ciba sold Galecron until 1988, when public pressure forced it to stop.[3]

Ciba executives, among the world's leading "corporate environmentalists," tend to dismiss these incidents as regrettable, a thing of the past. This snapshot focuses on the company's *recent* pesticide and agricultural biotechnology practices.

Monocrotophos in Ecuador

Researchers have determined that monocrotophos is mutagenic in laboratory animals and concluded that "pesticides containing monocrotophos as the active ingredient should be a great concern for human health."[4] Apparently, regulators in Switzerland, where monocrotophos's use is banned, agree. Yet Ciba continues to make monocrotophos at its plant in Monthey, Switzerland and export it to Ecuador where, as in many less-industrialized countries, monocrotophos causes regular poisonings.[5]

On average, one Ecuadorian farmworker is poisoned by monocrotophos every day.[6] In addition, according to the head of one village in Ecuador's Manabi Province, constant use of pesticides has contaminated water supplies. Pesticides have also poisoned many kinds of produce which in some cases the Ministry of Agriculture deemed too toxic to sell. The words of the people of Ecuador's Manabi Province testify to the dangers of using pesticides such as monocrotophos:

"Insecticides of the group phosphoric acid esters cause more than half of all the poisonings. Monocrotophos is one of the main products of this so-called OP (organophosphate) group. This year, monocrotophos has been the main agent of disease and death in a large number of poisoning cases."

– *Dr. Guido Teran, 1990* [7]

"The farmers have become utterly dependent on pesticides. Today, they can't grow anything without agrochemicals. The natural balance of the soil has been destroyed....But we are not just destroying our health and environment. We are also suffering economically from those agrochemicals....Our financial situation is deteriorating every day."
–   *Jorge Leon, farmer, 1990* [8]

"Ecuador is the trash can of the industrialized countries: banned for use in their countries of origin, pesticides are simply dumped in excess quantities here."
–   *Jose Toro, agronomist, 1990* [9]

DDT in Tanzania

Ecuador is not the only country in which Ciba has unloaded pesticides that are prohibited in Switzerland. In 1989-1990, the company, having decided to remove all products containing DDT from its inventory, formulated and shipped DDT-laden Ultracide combi to Tanzania.[10] DDT is banned in 29 countries, including Switzerland, and severely restricted in 23 others.

Ciba's shipment violated the UN Food and Agricultural Organization (FAO) Code of Conduct, a 1989 ruling by the Organization for Economic Cooperation and Development banning the sale of DDT-containing products in OECD countries, and internal policy guidelines Ciba itself established in 1990.[11] In March 1991, the Tanzanian subsidiary of Hoechst publicly revealed the shipment, pointed out the dangers Ultracide posed for people and animals, and charged Ciba with "unethical behaviour."[12]"Such behaviour," Hoechst noted, "does easily lead to the assumption that products are dumped in third world countries...."[13] Ciba quickly offered to buy back remaining stock of the pesticide and treat it as "toxic waste" in Switzerland, but as of fall 1992 the DDT still remained in Tanzania.

Atrazine in Europe

Ciba is the world's leading producer of atrazine, one of the group of triazine herbicides and a potential human carcinogen.[14] Environmentally persistent, atrazine has contaminated groundwater in many countries.

Forty-four thousand residents of one German city lost their drinking water due to atrazine contamination.[15] Atrazine and other triazine herbicides have contaminated streams and groundwater in 25 US states.[16] Several regions in Italy are considered "at risk" and 450,000 people are unable to drink their tap water because of atrazine.[17] It has been detected in rain water in Amsterdam.[18] As a result of these and other problems, most European countries have banned atrazine or intend to ban or restrict its use.[19]

Ciba defends atrazine – whose use in agrochemicals has accounted for about US$85 million in revenue annually – as the "aspirin of the agricultural industry."[20]

> "At a certain point the responsibility of the company ends."
> –   *Hans Geissbehler, spokesperson,*
>   *Ciba Agricultural Division, 1983* [21]

Patenting Life: Ciba and Biotechnology

> "Although biotechnology is undoubtedly capable of great benefit to mankind, it currently arouses widespread feelings of insecurity, rejection, and even resistance."
> –   *from Ciba's 1990 Annual Report* [22]

1. Herbicide Resistance

Biotechnology advocates say it will help supplement world food supplies and contribute to the development of more efficient food production.[23] Corporations, especially TNCs, are moving rapidly to privatize this field through patent applications.[24] The patenting of lifeforms will award the ownership of genes or natural processes to a handful of corporations.

For example, Ciba and Monsanto are designing and patenting crop seeds and plants that will resist their own herbicides, which as broad spectrum chemicals can kill crops as well as weeds. Eventually, these companies will be able to sell both the herbicides and the genetically-engineered plant seeds. As a result, the control these corporations have over farmers' autonomy and agricultural production throughout the world will increase. The costs of seed stock will also increase significantly.[25]

Ciba, among the world's top ten seed companies, ranks 3rd in plant-related patent applications to the European Patent Office.[26] Included are applications for plants that are resistant to atrazine and to sulfonylurea herbicides, one of which Ciba markets under the trade name Beacon.[27] Sulfonylureas are environmentally persistent, can cause heart and blood irregularities, and become more toxic in the presence of other chemicals.[28]

The global proliferation of herbicide-resistant plants will encourage greater herbicide use, and raise the likelihood of damage such use can cause. In addition, the release of genetically-engineered seeds and plants resistant to herbicides poses serious environmental risks. The herbicide-resistant plant may spread its genes via pollen to weeds, which would then be resistant to the herbicide. This risk is augmented if a weed is related to a crop, as is the case with rice, potatoes, and sugar beets. Disruption of ecosystems by releasing a genetically-engineered organism could result in a competitive advantage to the organism which would lead to the displacement of other species. Biological diversity is thus likely to be reduced. Herbicide-resistant plants may themselves become weeds if they survive as "volunteers" into the following crop.[29]

For the less-industrialized world, which is home to many major food crops' weedy relatives and whose great biological diversity is essential in maintaining disease and pest resistance, such risks are especially great. While less-industrialized countries will face these dangers, patent protection granted for lifeforms will tend to increase the trade and technology advantages of the Northern-based TNCs over the South.[30]

2. Poison-producing corn

Ciba has succeeded in transferring the genes from a bacterium, Bacillus thuringiensis (Bt), that is toxic to certain insects, into tobacco plants.[31] (Monsanto has done the same with cotton and tomatoes.) Under natural conditions, insects do not become resistant to Bt, a quality that has made the bacterium extremely attractive to the agricultural industry, and it has been applied as a biopesticide spray for years.[32] Currently, Ciba is pushing to develop Bt-engineered corn, which it will be able to patent and which it hopes to market by the late 1990s. Since maize seed is the biggest worldwide seed market of any field crop and the global market for corn is valued at some US$2.5 billion, the company considers these experiments to be "of crucial importance."[33]

It is not known if Bt-engineered plants are safe for human consumption.[34] Furthermore, in 1986 Bt-resistant insects were discovered in Hawaii, and have appeared subsequently in parts of Asia and the continental US.[35] Recently, scientists in the laboratory have created Bt-resistant insects in as little as twelve insect generations – less than five years. Widespread use of Bt-engineered plants, these scientists argue, will enlarge the numbers of insects resistant to the bacterium.[36] If this happens, Ciba and others will have spent many years and millions of dollars on a project with no long-term benefits to food production.

> []I believe that in the long-term the need for food production in developing countries...is providing a potential which in the long run we shall be able to supply."
> – T.W. Parton, Ciba, 1988 [37]

Ciba's Dirty Industry Movement

1. Swiss Opposition to Biotechnology

In its 1990 Annual Report, Ciba asserted, "Broadly based public discussion [about biotechnology] will be welcome, because it is the only way to arrive at a lasting social consensus on vital research questions."[38] In Switzerland, however, the discussion may have been too broadly based for Ciba. In 1992, impatient with delays in the approval procedure by the Swiss authorities and unwilling to fight environmental opponents, Ciba dropped plans to build a new biotechnology center, the Biotechnikum, in Basel, Switzerland. Instead, the company chose a site across the Rhine River in France, where Ciba believes there to be less opposition to genetic engineering.[39]

2. "Cancer Cluster" in New Jersey

Ciba brags that its Toms River facility in New Jersey won a 1990 National Environmental Achievement Prize from a "consortium of 22 national environmental protection organizations."[40] It is difficult to imagine what criteria were used in awarding that prize, given Ciba's long history of hazardous polluting at Toms River.

For over two decades, until 1992, the Toms River plant dumped over four million gallons a day of carcinogenic and teratogenic chemical waste into the Atlantic Ocean, 2500 feet off the shore of a popular beach and vacation community.[41] As early as 1969, Ciba also improperly landfilled chemicals that the company knew were hazardous at the Toms River site.[42] At least 1000 wells have been poisoned in the Toms River area, and residents have been forced to rely on bottled water.[43] The area, a national Superfund site, is known as a "cancer cluster" community, with 50-99 per cent higher rate of certain cancers, especially in children, than the average population.[44]

In 1992, two Ciba officials and the corporation pleaded guilty to illegally dumping waste at the Toms River facility. Ciba agreed to pay a total of US$61.35 million in fines, cleanup costs, and a donation to state conservation projects.[45] Two years earlier, Ciba had phased down its Toms River chemical manufacturing and moved some of the operations to a plant in San Gabriel, Louisiana where, according to local environmentalists, regulation is more lax.[46]

Endnotes

1. Special Report: Industry/activist dialogue, in *International Barometer,* Fall, 1988, p. S-12.
2. "Dangerous Exposures," in *India Today,* 15 January 1985, p. 88.
3. See Helen Howard, "Why a chemical firm sprayed Egyptian children with pesticide," in *New Scientist,* 10 February 1983, and "Dangerous Exposures." See also The Council on Economic Priorities (CEP) Corporate Environmental Data Clearinghouse, Environmental Profile of Ciba-Geigy Ltd. (Draft), 1991, p. 11.
4. M.F. Lin, C.L. Wu and T.C. Wang, "Pesticide clastogenicity in Chinese hamster ovary cells," in *Mutation Research,* 188(1987), p. 248.
5. See Pesticide Action Network (PAN), "The FAO Code: Missing Ingredients," prepared by the Pesticides Trust for PAN, Pesticides Trust, London, 1989, as cited in Greenpeace, "Dangerous Pesticide Briefing," vol. 3, no. 1, 1991, p. 3.

6. Presentation by Mercedes Bolanos, National Coalition Against the Misuse of Pesticides, Annual Pesticide Forum, 23-26 March 1990.

7. From Greenpeace video "Pesticides Made in Switzerland," 1990.

8. Ibid.

9. Ibid.

10. "C-G to repurchase Ultracide combi lot," in *Agrow,* 13 September 1991.

11. Ibid. See also "C-G admits contravention of FAO code," in *Agrow,* 14 June 1991, and "DDT-Lieferung nach Tansania: Ciba-Geigy zieht Konsequenzen," in *Basle-Zeitung,* 24 May 1991.

12. 19 March 1991 letter of Hoechst Tanzania Ltd. to Ciba-Geigy Trading and Marketing Services Co., Ltd.

13. Ibid.

14. CEP, op cit, p. 10.

15. "High atrazine levels found in FRG drinking water," in *Agrow,* 16 March 1990.

16. *Chicago Tribune,* 2 June 1991.

17. Greenpeace, "Dangerous Pesticide Briefing," p. 5.

18. *Agrow,* 10 August 1988, p. 6.

19. "Dangerous Pesticide Briefing," p. 5.

20. CEP, op cit, pp. 4-5.

21. "Why a chemical firm sprayed Egyptian children with pesticide," op cit.

22. 1990 Annual Report of Ciba-Geigy, p. 12.

23. See, for example, "Transgenic maize," in *Ciba-Geigy Journal,* March 1990, or the International Chamber of Commerce's Business Brief #8 on Biotechnology, 1992.

24. Henk Hobbelink, *Biotechnology and the Future of World Agriculture,* London, Zed Books, 1991, pp. 115-116.

25. CEP, op cit, p. 9.

26. Hobbelink, pp. 115-116.

27. CEP, op cit, p. 3, "Sulfonylureas," *chemicalWatch,* March 1991.

28. "Sulfonylureas", ibid.

29. For more information on the potential problems and dangers of herbicide-resistance plants see: N. Ellstrand, "Pollen as a vehicle for the escape of engineered genes?", *TIBECH,* 6:S30-31; M. Williamson, in "Herbicide resistance in weeds and crops: 11th Long Ashton International Symposium," J.C. Caseley, G.W. Cussans, M.S. Kemp, J.R. Moss, & R.K. Atkin(eds), 1990; Hobbelink, op. cit.; V. Shiva, "Biodiversity, biotechnology, and profit," *The Ecologist,* 20:44-47; H. Darmency & J. Gasquez, "Fate of herbicide resistance genes in weeds," in *Managing resistance to*

agrochemicals: from fundamental research to practical strategies," M.B. Green, H.M. LeBaron, & W.K. Moberg(eds), 1990; P. Longden, "Weed Beet"84, *Beet Review,* 52:77; M. Williamson, J. Perrins, & A. Fitter, "Releasing genetically engineered plants: present proposals and possible hazards," *TREE,* 5:417-419; and H.M. LeBaron & J. McFarland, "Herbicide resistance in weeds and crops," in *Managing Resistance to Agrochemicals,* eds. M.B. Green et al, pp. 326-352.

30. Hobbelink, op cit, p. 119.

31. "Transgenic maize," op cit.

32. "Squandered Weapons," in *leafLET The Newsletter for Seed Savers,* vol. 1, issue 1, winter 1992, and Christopher Anderson, "Researchers ask for help to save key biopesticide," in *Nature,* 20 February 1992.

33. "Transgenic maize", op cit, and CEP, op cit, p. 10.

34. Dr. Rebecca Goldberg, "Are B.t.k. plants safe to eat?" in "Global Pesticide Campaigner," January 1991, as cited in CEP, p. 6.

35. "Squandered Weapons," op cit.

36. Anderson, op. cit.

37. *International Barometer,* op cit.

38. Annual Report 1990, p. 13.

39. "C-G to build Biotechnikum in France," in *Agrow,* 17 January 1992.

40. 1990 Annual Report of Ciba-Geigy, p. 14.

41. Alexander Cockburn, "Pollution in New Jersey: The Way the World Really Is," in *The Wall Street Journal,* 28 July 1988, and "Kean Gets Bill Banning Ocean Dumping," in The *Philadelphia Inquirer,* 23 May 1989.

42. "Ciba-Geigy Arraignment Set," *The New York Times*, 2 November 1985, and "Probe Finds Ciba was U.S. Dump – Wastes from 57 Sites," in *The Toms River(N.J.) Observer,* 10 July 1987.

43. New Jersey School of Medicine Study, on file with Frank Livelli of S.O.O.S., Lavallette, NJ.

44. Cockburn, op cit.

45. Elisabeth Kirschner, "Ciba-Geigy and New Jersey settle Toms River battle," in *Chemicalweek,* 11 March 1992, p. 14.

46. Ciba-Geigy Corporation (US), "Report 90," p. 2. Cockburn, op cit. Also personal communication of Jed Greer with the Louisiana Environmental Action Network.

Source: From information packet on GM's Geo Tree Planting Program.

Greenwash Snapshot #12

GENERAL MOTORS CORPORATION

A case study in auto-dependency, lost jobs, and pollution.

General Motors Corporation (GM)
Chairman: John F. Smith, Jr.
Headquarters: 3044 W. Grand Boulevard Detroit, Michigan 48202, USA
Tel: 313-556-5000   Fax: 313-556-5108
Products: motor vehicles.

GM has facilities at 283 US locations and in 42 other countries. It is a signer of the ICC Rotterdam Charter.

> "At General Motors, we recognize the effects that cars and their manufacture have on the environment. We understand the relationship better than any other car maker in the world."
> – *GM Earth Day 1990 advertisement*

GM, Cars, and the Environment

General Motors ought to understand the relationship between cars and the environment; it is the world's number one manufacturer of motor vehicles, and motor vehicles are in turn the world's number one source of air pollution. The world's 550 million cars, trucks, and commercial vehicles consume one-third of the world's oil.[1] General Motors vehicles release an estimated two per cent of the carbon dioxide emitted into the air each year. In the Organization for Economic Cooperation and Development (OECD) countries, GM accounts for an estimated 11 per cent of the carbon monoxide, eight per cent of the nitrogen oxides, and six per cent of the hydrocarbons emitted by vehicles annually.[2] The American

Lung Association calculates that the health costs attributable to vehicle emissions in the US alone are US$25 billion per year.[3]

To help greenwash the harmful consequences of its automobile production and use, GM has a program of planting a tree regionally for each car sold in its Geo division (at the participating dealerships). Purchasers of a Geo vehicle receive fancy kits with Geo Tree logo lapel pins and certificates from the US Forest Service.[4]

But the positive contribution of this program is minimal. To counteract the amount of carbon dioxide produced by a single Geo car, GM would have to plant 734 trees over the estimated ten-year lifetime of each vehicle.[5] Furthermore, planting trees does not mitigate the impact of other automobile pollutants such as reactive hydrocarbons and nitrogen oxides, which are among the main components of urban smog.

If GM really acknowledged the effects of their products on the environment and was committed to more than greenwash, we might expect to see more emphasis on fuel efficiency, renewable fuels, and even public transportation and bicycles.

But GM is a staunch opponent of raising fuel economy standards. Along with other automakers, GM has consistently, and successfully, pressured the US government to roll back fuel economy standards and reaped huge profits as a result. In 1986, the National Highway Traffic Safety Administration (NHTSA) would have fined GM US$385 million for failing to meet the 1985 corporate average fuel economy (CAFE) standard of 27.5 miles per gallon (mpg) but for the rollbacks given that year.[6]

GM has lobbied against fuel efficiency and has sponsored political action committees (PACs), which give campaign contributions to US political candidates, to oppose tighter emissions standards.[7]

Meanwhile, a solution to GM's well-publicized economic problems (in 1991 the company recorded the largest loss in the history of US business) – and a benefit for the environment – is right under its nose. GM has demonstrated a prototype car, the Ultralite, which gets 100 miles per gallon of gasoline. Such a high-efficiency vehicle could be part of the future of private automobiles and a potentially huge seller for GM, but the company has no plans to put the prototype into production.[8] Moreover, only in 1996 did GM announce it would begin selling a very small number of electric cars in the Western US states of California and Arizona, six years after the company first showed the car to the public. Ironically, GM's announcement came just several weeks after California had bowed to the automaking industry's pressure and abandoned a mandate that two per cent of all cars sold in the state be electrically powered by 1998. That mandate would have required GM and other automakers to market tens of thousands of electric vehicles in California by the end of the decade.[9] GM will not say how much money it has invested in electric vehicle research.[10]

Spreading Car Dependency

During the first half of this century, GM and other large corporations bought up rail companies and dismantled public rail transport in 45 US cities in order to increase the demand for private vehicles.[11] The environmental and social effects of American car dependency are well known. The US and Canada are the largest per capita users of gasoline in the world, and cars continue to be both the fastest growing energy demand and the fastest growing air pollution source in North America.

In the second half of this century, GM and other automakers have steadily expanded into Latin America where GM is the largest US vehicle manufacturer. Road and highway construction in the region has diverted funds from public transportation in cities where the majority of the population cannot afford a car. In Brazil, it is estimated that in order for the car market to expand the income of the top ten per cent of the population would have to increase, worsening the gap between rich and poor.[12]

The effect that non-existent or inadequate public transit has on the majority of the urban poor is profound, controlling their access to jobs, health care, and basic services. Commutes of two hours or more in each direction are routine for people who come from the shantytowns on the outskirts of Sao Paulo and other booming Latin American cities into the center to work.[13]

Unsustainable Jobs, Lasting Pollution

GM has based its opposition to tighter emission rules on the need to preserve US jobs. In 1988, former GM head Robert Stempel asserted: "If stringent new proposals are put into effect...it could lead to the curtail-

ment of auto production here in Michigan and around the country...."[14] But
this assertion distracts from an important reason GM's jobs in the US are
unsustainable. The company is striving to become the largest US vehicle
manufacturer outside the United States and pledges to "knock the hell out
of the competition."[15]

To cut costs, the company is laying off workers in the US – 30,000
in the late 1980s and a planned 75,000, or 18 per cent of its workforce, by
the mid-1990s. Facilities are being moved to Mexico and other countries
where wages are a fraction of those in the US.[16] GM's movement has
come in spite of economic incentives offered by communities and wage
and benefit concessions which have been given by autoworkers in the US
throughout the late 1980s and 1990s. In North Tarrytown, New York,
where years of concessions given by the town and union failed to prevent
closure of the local GM plant, one official observed: "We're left holding
the bag, and the bag is empty."[17]

In a notable challenge to corporate greed, a county circuit court judge
temporarily prevented GM from closing its Willow Run plant in Ypsilanti,
Michigan because GM had promised to keep the plant running in return
for tax abatements of US$13.5 million granted in 1984 and 1988.[18] In late
1992, however, GM announced it was closing the facility, adding an extra
shift at a plant in Texas, and cutting 4,500 jobs at Willow Run.[19]

The township took GM to court. In February 1993, the county judge
ruled in Ypsilanti's favor, placing the rights of the community over those
of the company. "There would be a gross inequity and patent unfairness,"
Judge Donald Shelton asserted, "if General Motors, having lulled the
people of the Ypsilanti area into giving up millions of tax dollars which
they so desperately need to educate their children and provide basic
governmental services, is allowed to simply decide that it will desert
4,500 workers and their families because it thinks it can make these same
cars a little cheaper somewhere else."[20]

In August 1993, a Michigan appeals court reversed the February
ruling, and a month later the state's Supreme Court refused to hear an
appeal from the appeals court decision.[21] According to the attorney who
represented Ypsilanti, the failure of the higher courts to side with the
townspeople "allows corporations like GM to continue to practice deceit
and dishonesty and to bamboozle communities when they seek tax sub-

sidies in exchange for creation of employment."[22]

Communities which do become the sight of a GM plant must be concerned about environmental contamination. Wastewater discharges from a GM plant in Matamoros, Mexico were found in 1990 to have extremely high levels – 6,600 times US standards – of the toxic solvent xylene, which can cause respiratory irritation, amnesia, brain hemorrhages, internal bleeding, and damage to the lung, liver, and kidneys. The discharges went into an agricultural drainage canal that leads to the Rio Grande, a source of drinking water.[23]

> "I work at Rimir, a General Motors plant....I am very concerned about the contamination caused by toxic chemicals that come out of the Rimir plant. At the Rimir plant we paint automobile bumpers. To clean paint guns and paint lines we use chemical solvents....We run the solvent through the guns and lines to purge the paint....All of the liquid solvents go down into the floor drain and into a pipe that leads to a canal on the side of the Rimir plant."
> – *anonymous GM worker, 1990* [24]

> "We are very proud of our operations in Mexico."
> – *Robert Stempel, former GM Chairman*
> *& Chief Executive Officer, 1991*[25]

In 1990, the US Environmental Protection Agency (EPA) issued a US$78 million clean up order to GM for a foundry in Massena, New York. The facility and surrounding areas had been contaminated with the equivalent of 55,000 truckloads of PCBs and other toxic chemicals. In 1991, the EPA fined GM and two New York state companies US$35.4 million for improperly disposing of PCB-contaminated sludge there. This was one of the largest penalties ever levied by the EPA. In 1990, GM paid the Occupational Health and Safety Administration US$360,000 for alleged violations of health and safety rules at US plants. In 1987, GM paid US$500,000 for similar violations.[26]

Millions of GM products have violated US vehicle emissions standards. From 1982 to 1990 GM voluntarily recalled or was ordered to recall 7.5 million cars. Over two-thirds were for excessively high emissions of nitrogen oxides, carbon monoxide, and hydrocarbons; the rest were for faulty emission control systems.

US EPA recall orders or GM voluntary recalls from 1982-1990 include: 2,902,000 vehicles for excessive emissions of nitrogen oxides; 1,178,000 vehicles for defective catalytic converters; 1,171,843 vehicles for excessive hydrocarbon emissions; 1,160,000 vehicles for defective emission control systems; 500,000 vehicles for excessive carbon monoxide and evaporative emissions; and 598,588 vehicles for excessive evaporative emissions.[27]

"For more than three decades, GM has seen a clean and healthy environment as a top priority. We take pride in our leadership role in reducing emissions from both vehicles and plants and in our work to minimize wastes and to dispose of those wastes in an environmentally sound manner."
–   Former GM Chairman Stempel,
    GM Annual Report, 1991[28]

Motor Vehicle Air Pollution		
Vehicle Exhaust Pollutant	**Environmental Effects**	**Health Effects**
carbon monoxide	Helps increase the buildup of methane, an important greenhouse gas.	Lethal in large doses; affects central nervous system; aggravates heart disorders; impairs oxygen carrying capacity of blood.
nitrogen oxides	Acid rain, contributes to buildup of ground-level ozone, a greenhouse gas 2,000 times as effective as carbon dioxide in retaining earth's heat.	Irritate or impair respiration; lessen resistance to infection. (Ozone causes eye, nose and throat irritation and can damage vegetation.)
hydrocarbons	Contributes to build-up of ground-level ozone.	Drowsiness; coughing; eye irritation.
Other toxic vehicle emissions include benzene, aldehydes, and lead. Adapted from Greenpeace International report *The Environmental Impact of the Car,* p. 32.		

Endnotes

1. Greenpeace, *The Environmental Impact of the Car,* Greenpeace International, Amsterdam, 1991. pp. 5-6.
2. Data from *The Environmental Impact of the Car,* p. 11, pp. 19 & 21. Based on GM's worldwide #1 ranking among motor vehicle manufacturers in 1988. That year, GM's market share was 16 per cent which when multiplied by 14 per cent–the amount of global atmospheric carbon dioxide attributable to tailpipe emissions alone – equals over two per cent. To get figures for OECD countries, that same market share is multiplied by 47 per cent (nitrogen oxides), 39 per cent (hydrocarbons), and 66 per cent (carbon monoxide).
3. James Cannon, spokesperson for American Lung Association, 19 January 1990.
4. Bruce Horovitz, "Honk if You Love the Environment," *Los Angeles Times,* 23 March 1993. Information about the tree-planting program from personal communication with representative of the Pearlman Group, which created the program for Geo.
5. Horovitz, ibid.
6. *Environment Reporter,* 4 October 1985, 10 October 1986.
7. *The Environmental Impact of the Car,* op cit, p. 54.
8. Greenpeace, "Fact Sheet on Auto Manufacturer Greenwash," 1993.
9. Lawrence Fisher, "GM, in a First, Will Sell a Car Designed for Electric Power This Fall," *The New York Times*, 5 January 1996.
10. Ibid.
11. *The Environmental Impact of the Car*, op cit, pp. 53-54.
12. Rhys Jenkins, *Transnational Corporations and the Latin American Automobile Industry,* The Macmillan Press, London, 1987. p. 5, p. 105, pp. 239-243.
13. Marcia Lowe, "Shaping Cities: The Environmental and Human Dimensions," Worldwatch Institute Paper #105, October 1991.
14. "GM President Calls for Reduced CAFE Standards to Save Jobs," PRNewswire, 14 September 1988.
15. Jenkins, op cit, p. 247.
16. Doron Levin, "G.M. Retrenchment to Shut 21 Plants, Losing 70,000 Jobs," in *The New York Times*, 19 December 1991.
17. Thomas Lueck, "Business Incentives: A High-Priced Letdown," *The New York Times*, 8 March 1992.

18. John Borsos, "The Judge Who Stood Up to G.M.," in *The Nation*, 12 April 1993, pp. 488 & 490. Also "Stopping Capital Flight – A Strategy for Corporate Accountability," interview with Douglas Winters, in *Multinational Monitor*, June 1993, p. 12.

19. Boros, ibid.

20. Quoted in Borsos, ibid.

21. Donald Levin, "Court Backs GM on Plant Closings," in *The New York Times*, 5 August 1993, and Aaron Freeman, "GM Abandons Ypsilanti," in *Multinational Monitor*, September 1993, p. 4.

22. Quoted in Freeman, ibid.

23. National Toxics Campaign, *Border Trouble: Rivers in Peril – A Report on Water Pollution Due to Industrial Development in Northern Mexico*, Boston, 1991. pp. 4, 9.

24. Quoted in *Border Trouble*, p. 12.

25. BNA Environment Daily, 3 June 1991.

26. See: United Press International, 19 December 1990, 18 March 1991; *Environment Reporter*, 10 October 1986; United Press International, 21 November 1990.

27. Record of GM recalls and recall orders in *Environment Reporter* on relevant dates or in United Press International.

28. BNA Environment Daily, 3 June 1991.

Earth, Air, and

Performance sells cars. Performance, image and style. Yes, people want to conserve energy and safeguard the environment. And laws are beginning to insist on it. But the car buyer has to have—so the carmaker has to deliver—good acceleration, handling, cornering, braking, space, comfort and the feel of a winner. In a word, performance. ¶ Is automotive design on a collision course with environmental reality? Or is it possible that the

Source: Alcoa 1991 Annual Report.

Greenwash Snapshot #13

ALUMINUM COMPANY OF AMERICA

A case study in the environmental and social costs of aluminum production.

Aluminum Company of America (Alcoa)
Chairman & Chief Executive Officer: Paul O'Neill
Headquarters: 1501 Alcoa Building Pittsburgh, Pennsylvania 15219 USA
Tel: 412-553-4545   Fax: 412-553-4498
Major business: aluminum for packaging, transportation, building, and industrial markets.
Major subsidiaries: Alcoa of Australia; Alcoa Aluminio SA(Brazil)

Alcoa has 159 operating and sales locations in 22 countries. It is a member of the WBCSD.

> "In the words of one environmentalist within our ranks: 'If it moves, it ought to be aluminum.'"
> – *from Alcoa's 1991 Annual Report*[1]

A glossy photograph of a car adorns the cover of Alcoa's 1991 Annual Report. Inside, the company devotes seventeen pages to more photos and text which advertise the ways greater aluminum use in cars will save energy, conserve natural resources, and reduce air pollution. As the world's largest aluminum producer, Alcoa has a clear economic motive in depicting aluminum as a savior of the environment. To the extent that Alcoa's overriding emphasis on aluminum in cars perpetuates and increases the world's dependency on the motor vehicle industry, how-ever, it does more harm than good. Cars, trucks, and commercial vehicles are the biggest sources of global atmospheric pollution, consume one-third of the world's oil, and require vast amounts of energy to manufacture (see GM case study).[2]

In countries such as Brazil, the aluminum industry's tremendous energy demands have led to the construction of habitat-destroying hydroelectric dams and have added to an ongoing fiscal crisis in the country's electricity sector. Alcoa's aluminum operations have also forced the eviction of 20,000 people on Brazil's Sao Luis Island to make way for plant expansion and contributed to PCB contamination of rivers near a Mohawk Nation reservation in North America.

Environmental and Economic Costs of Aluminum Production

The production of aluminum itself carries a high ecological and economic price. At every stage, aluminum production degrades and pollutes the environment. The strip-mining of bauxite, aluminum's principal ore, probably destroys more of the earth's surface area than the mining of any other metal. The extraction of aluminum from bauxite generates an equivalent amount of toxic waste called "red mud" that is rich in metallic oxides and other contaminants. This toxic mud is frequently left near mines which inevitably leads to surface and groundwater contamination.

Aluminum smelters are an important source of numerous air pollutants including aluminum dust, fluoride dust and gases, hydrocarbons, carbon monoxide and sulfur dioxide, which contributes significantly to acid rain.[3]

Alcoa claims that if all cars in the US used aluminum for parts that are currently available, fuel savings would reduce annual carbon dioxide emissions – the main cause of global warming – by 98 million tons.[4] Recently, however, a researcher has discovered that primary aluminum smelting is the only known major human-caused source of CFC-14 and CFC-116, perhaps the most potent greenhouse gases being emitted in large amounts. Scientists estimate that these gases will stay in the atmosphere for 10,000 years and are equal to the greenhouse contribution of between 15 and 20 tons of carbon dioxide per ton of aluminum.

This means that in 1990, when 18 million tons of aluminum were produced, the aluminum industry emitted the equivalent of between 270 and 360 million tons of carbon dioxide. Future reports of the Intergovernmental Panel on Climate Change will include CFC-14 and CFC-116 from aluminum smelting in their assessments of major greenhouse gases.[5]

Aluminum production is also one of the more energy-intensive industries on earth.[6] Each year, it consumes at least 250 billion kilowatt hours of electricity, or about one per cent of global energy supply. In 1990, the world aluminum industry needed almost as much electricity just to convert alumina into aluminum as was used in all of Africa.[7] Because of this, many countries give generous energy subsidies to aluminum producers which allow the companies to pay less for power than other consumers.[8]

These subsidies often obscure the reality of the economic benefits generated by the aluminum industry. In Brazil, for example, the amount of energy used by aluminum smelting doubled between 1982 and 1988, when the industry consumed 12 per cent of the country's electricity. Nearly all the increase went to produce aluminum for export. Although such exports provide foreign exchange, a government research institute in Brazil calculates that subsidies to aluminum producers comprise 60 per cent of the metal's export value.[9] In 1989 alone, this amounted to some US$600 million.[10]

Energy prices in Latin America, Africa, and Asia are one-half to one-seventh those in the US or Western Europe. This disparity has encouraged transnational aluminum corporations to relocate their smelting and manufacturing operations to these regions.[11] The less-industrialized world's share of global production of aluminum has doubled consistently every ten years since 1960 and is expected to reach nearly 50 per cent by the year 2000. Alcoa now has more operations for primary aluminum and fabricated aluminum products in Brazil (nine) and Mexico (six) than any other country except the US (which has 26).[12]

This shift can wreak economic havoc with electric power systems in less-industrialized countries. Aluminum production has caused a dramatic increase in Brazil's energy use. In turn, this has contributed to an ongoing fiscal crisis in the country's electricity sector.[13] And in some areas, the consequences of this dirty industry movement have been truly devastating.

Aluminum and Hydropower: Alcoa and Destruction in Brazil

Approximately 40 per cent of the electricity used by the aluminum industry comes from hydropower – enormous dam projects that in the Americas and Africa have been ruinous to the environment and have dislocated thousands of people. In some cases the dams themselves were built for the vast electricity needs of aluminum smelters.[14]

Built by Brazil's state-run power company, the huge Tucurui dam on the Tocantins River in the Amazon basin exemplifies the harm such projects can cause. About one-third of the dam's electricity goes to aluminum producers, among them the world's second largest combined aluminum refinery and smelter on Sao Luis Island owned by Consorcio Alumar, a joint venture in which Alcoa is the majority partner (Shell–Billiton is the other partner). The dam's construction flooded 243,000 hectares, including six towns and two Indian reserves. It has threatened the habitat and existence of many plants and animals – especially marine animals such as fish, dolphins, turtles, manatees, and caimans – through contamination and deforestation.[15]

The decline in fishstocks due to the hydroelectric project jeopardized the livelihood of nearby fishing villages. Malaria has increased in the area since the dam was built.[16] At least 20,000 people were evicted from their homes on the island of Sao Luis to make way for Alumar's production facilities as well as a railway station.[17] In addition to the railway, roads and power transmission lines pass through or near twelve of the region's tribal communities, many of which now compete for food and land with a growing number of peasant squatters attracted by the transportation network.[18] ·

"Alcoa is the devil itself, and I'm very afraid of them, equally from the pollution that comes from below as well as that from above. Alcoa produces poison which enter the air and the ground. When there's a lot of smoke, if there's no rain, it falls with the dew. So, it's destroying everything....Their waste pond...is close to where we work. The river...was affected by their waste. Today it's just a trickle. Their waste flowed down and killed everything. The water is polluted and we have to drink well water."
–   *Livia Silves Valdes, resident of the Igarau community,*
   which resisted relocation by Alcoa, 1987[19]

"Alumar is a plant that operates in a rigorously clean manner. The
residue of the bauxite refining are placed in impermeable bags within
the pond....The ponds are a masterpiece of engineering...."
– *José De Jesus Brito, Alumar spokesperson 1987* [20]

Alcoa, Aluminum, and Hazardous Waste

Alcoa has also poisoned the North American environment, via its
aluminum smelting and fabrication plant in Massena, New York.
The contamination is worst near the St. Regis Mohawk Nation Reservation – which the Mohawks call Akwesasne – along the St. Lawrence
River. In 1989, the US Environmental Protection Agency (EPA) `declared
rivers near the reservation a hazardous waste site. The EPA ordered Alcoa
and Reynolds Metals Company to clean up large amounts of
polychlorinated biphenyls (PCBs) which they had discharged for
many years.[21] (In 1990, the EPA also issued an order to General
Motors to clean up PCBs in the Massena area.)

The EPA move followed a state Health Department study which
showed that PCBs were accumulating in the breast milk of Mohawk
mothers. In addition, according to a local environmentalist, the PCBs
"effectively destroyed the fishing industry" and contaminated farmland
on the reservation, while high fluoride levels caused bone deterioration
in cattle.[22] Pollution from these PCBs and other hazardous chemicals is
responsible for the disease, premature death, and declining birth rate in
the St. Lawrence's white beluga whale population. One expert calls the
belugas "probably the most contaminated mammal in the St. Lawrence
River ecosystem."[23]

In 1991, Alcoa agreed to pay a criminal fine of US$3.75 million and
a civil penalty for US$3.75 million for a number of NY state environ-
mental offenses at its Massena plant.[24] Alcoa was charged with improperly handling PCBs and with the illegal disposal of acidic compounds
used in an aluminum-cleaning process. Since 1983, Alcoa had poured
these compounds down a sewer hole where they mixed with wastewater
that drained into a tributary of the St. Lawrence. The criminal fine
was the largest amount ever levied for a hazardous waste violation
in the US.

"If I didn't believe my children and grandchildren will be able to swim in the St. Lawrence then there is no reason for me to live."
–   *Henry Lickers, Director, St. Regis*
    Environmental Division, 1987 [25]

"Our traditions survive in doing things the Mohawk way. Our whole ceremonial life, our cosmological life, is based on nature. Without that river, we lose Akwesasne."
–   *Katsi Cook, Mohawk midwife, 1988* [26]

In 1989, two Alcoa alumina plants and one smelter were among the top ten sources of hazardous toxic pollution in the National Wildlife Federation's "Toxic 500" list; one plant, at Point Comfort, Texas, was ranked #1. Alcoa claimed that under EPA rule changes, none of the plants would appear on the list.[27]

The Limits of Aluminum Recycling

In the US, almost one-third of aluminum is used in packaging, mostly beverage cans. Over half of aluminum cans in the US are recycled, and Alcoa, the leading supplier of aluminum sheet for beverage cans, has aggressively publicized its recycling efforts for these containers.[28] None-theless, the quantity of aluminum thrown away in the form of beverage cans in the US is greater than the total use of aluminum by all but seven nations.[29] Moreover, even 100 per cent recycling of aluminum cans would not be efficient compared to some alternatives, such as refillable containers.[30]

The primary goal of recycling – to reduce demand for raw material – is not promoted by Alcoa, nor is aluminum recycling leading toward this goal (primary aluminum production continues to increase). In itself, more recycling will not resolve the environmental and economic problems associated with aluminum production. With per capita aluminum consumption in the US at around 42 pounds annually (as opposed to under two pounds in China and Mexico, for example), Alcoa and other aluminum producers should focus their efforts on the fundamental goal of reducing production and consumption of this environmentally dangerous metal, rather than on greenwashing the use of even more aluminum.

Endnotes

1. Alcoa 1991 Special Report: Aluminum and the Future of Transportation.
2. *The Environmental Impact of the Car,* Greenpeace International, Amsterdam, 1991, especially pp. 5-6.
3. John Young, "Aluminum's Real Tab," in *World Watch,* March-April 1992, p. 27 and John Young, "Mining the Earth," in *State of the World 1992* A Worldwatch Institute Report on Progress Toward a Sustainable Society, New York, 1992, p. 107. See also "Fear surrounds RTZ aluminum plant," in *Labor Research,* June 1986, p. 15.
4. Alcoa 1991, p. 24.
5. Dean Abrahamson, "Sources and Sinks of Greenhouse Gases in Sweden: A Case Study," in *Ambio,* vol. 21, no. 2, April 1992, and "Greenhouse Gas Emissions from Aluminum Production An IPCC Oversight," submitted to *Nature,* 30 December 1991. Also personal communication of Jed Greer with author.
6. See Young, "Aluminum's Real Tab," p. 28, "Mining the Earth," pp. 109-110, and also United Nations Centre on Transnational Corporations (UNCTC), *Climate Change and Transnational Corporations Analysis and Trends,* Environment Series No. 2, New York, 1992, pp. 76-77.
7. Young, "Aluminum's Real Tab," p. 26.
8. Ibid, pp. 29-30, and UNCTC, p. 78.
9. Ibid, Young, p. 32.
10. According to *Mining Journal,* 30 March 1990, p. 259. Brazil earned US$1 billion from aluminum exports in 1989.
11. Rhys Jenkins, *Transnational Corporations and Uneven Development,* Methuen, London, 1987, pp. 108-109.
12. Alcoa 1991, p. 5.
13. Young, "Aluminum's Real Tab," p. 32.
14. Ibid, p. 26, and UNCTC, op cit, p. 77.
15. Ibid, Young, p. 31, and UNCTC, op cit, p. 78.
16. UNCTC, op cit, p. 78.
17. Dave Treece, "Brutality and Brazil: The Human Cost of Cheap Steel," in *Cultural Survival Quarterly,* 13(1), p. 30.
18. Ibid, p. 30.
19. From film "AMAZONIA: Voices From the Rainforest," Amazonia Films, PO Box 77438, San Francisco, CA 94107.
20. Ibid.

21. "EPA Reportedly to Declare Mohawk Reservation Rivers Hazardous Waste Sites St. Regis Indian Reservation," UPI, 5 October 1989.

22. Tom Spears, "Sickly St. Lawrence Gets Daily Dose of Poison; Wasting a River," in the *Ottawa Citizen*, 8 April 1990. See also Mike Thompson, "Pollution remains a big problem," *The Standard Freeholder,* 12 November 1987.

23. Biologist Daniel Green of the Montreal-based Society to Conquer Pollution, in conversation with Jed Greer. Also "St. Lawrence River Kills Beluga Whales," in *The New York Times*, 12 January 1988.

24. Elizabeth Edwardsen, "Alcoa Pollution," AP, 11 July 1991.

25. Quoted in Mike Thompson, "Pollution remains a big problem,"op cit.

26. Quoted in "Mohawk: Toxic Waste, Contaminated Animals Threaten the Culture," in *The Los Angeles Times*, 24 January 1988.

27. "Firms Contest Toxic Ratings," AP, 14 August 1989.

28. "Alcoa Sponsors Recycling Ads," American Metal Market, 6 November 1990.

29. Young, "Aluminum's Real Tab," p. 33.

30. Ibid, p. 33.

Source: Cartoon by Kirk Anderson.

Source: *For Gaia The Environmental Affairs Department at Mitsubishi Corporation,* 1991. Mitsubishi Corporation provided the funding and data for the comic.

Greenwash Snapshot #14

MITSUBISHI GROUP

A case study in tropical deforestation.

Mitsubishi Group (Mitsubishi)
Chairman: Shinroku Morohashi
Headquarters: Mitsubishi Shoji Kabushiki Kaisha, 6-3 Marunouchi
2-chome, Chiyoda-ku, Tokyo 100-86, Japan
Tel: 81-3-3210-2121 Fax: 81-2-3287-1321
Major businesses: metals; fuel products; machinery; chemicals; electrical
equipment; timber.

Ordered split up by the US after World War II, Mitsubishi Group consists
of about 160 separate companies with interlocking ownership. Unless
otherwise noted, here Mitsubishi refers to Mitsubishi Corporation, the
Group's trading company, which has 55 offices in Japan and 107 in other
countries. Other important Mitsubishi companies are Mitsubishi Heavy
Industries, Mitsubishi Electric, Mitsubishi Bank, and Mitsubishi Kasei
(Chemical). Mitsubishi is a member of the WBCSD.

Mitsubishi Group's businesses range from chemicals to nuclear technology to finance. Mitsubishi Kasei is Japan's largest maker of polyvinyl
chloride (PVC), with plants in dozens of countries. A Malaysian high
court ordered a Kasei subsidiary to shut down temporarily after local
people sued the company for gross negligence in the storage and handling
of radioactive waste from its mining operation. Mitsubishi Heavy Industries and other Mitsubishi companies and joint ventures have built
fourteen nuclear reactors in Japan, sold nuclear technology to China, and
are bidding on nuclear reactors for Indonesia. As of early 1992, Mitsubishi
Bank held 186.3 billion yen in debt from "less-developed countries."[1]

But it is the logging operations of this giant complex of corporations which have generated outrage around the world.

Mitsubishi and Logging: Comic Book Greenwash

"In practical terms, no commercial logging of tropical moist forests has proven to be sustainable from the standpoint of the forest ecosystem, and any such logging must be recognized as mining, not sustaining, the basic forest resource."
–   *L. Talbot, in World Bank report, 1990* [2]

The world's rainforests and the life they support are in dire jeopardy. UN Food and Agricultural Organization (FAO) studies indicate that depletion of rainforests is between 11 million and 17 million hectares each year. Of this total, 59 per cent is located in Central and South America, 32 per cent in Africa, and 18 per cent in Asia. Although tropical forests cover only seven per cent of the earth's land surface, they contain more than half of all living species.[3]

Mitsubishi, and its affiliate Meiwa Trading Company, is a leading destroyer of tropical (and non-tropical) forests, with operations in Chile, Venezuela, Bolivia, Brazil, the Philippines, Canada, Papua New Guinea, Malaysian Borneo, and Siberia. The Malaysian operation, in Sarawak State, is the most notorious. Mitsubishi owns 60 per cent of Daiya Malaysia Sdn. Bhd. (including private Japanese investors of up to 19 per cent), which has been logging its 90,000 hectare timber concession at a rate which will eliminate the area's forests within 12 years.[4]

Rapid deforestation and destruction of their homelands have fueled a strong protest movement against forest cutting by indigenous peoples of Sarawak such as the Penan, Kayan, Iban, Kenyah, and Kelabit groups, as well as an international campaign including a boycott of Mitsubishi Group consumer products. Those protests have not ended Mitsubishi's destructive role in Sarawak, but instead have inspired some bizarre greenwash by the company.

Mitsubishi Corporation's Environmental Affairs Department was immortalized in a comic book it funded and aimed at Japanese high school students. The comic follows the career of a middle-level executive named Hino, who travels around the world to find out the truth about Mitsubishi and rainforests after reading criticism of the company's practices. Not surprisingly, Hino ends up believing that shifting cultivation and poverty are the true cause of most deforestation, that local people want this kind of development anyway, and that Mitsubishi's critics are engaging in Japan-bashing.

After coming under fire from environmentalists in Japan and internationally, the comic book was withdrawn from circulation by Japan's Department of Education in March 1992.

Mitsubishi's Myths

In more serious fora, Mitsubishi has put forth similarly misleading arguments about its logging activities. The company says that "most deforestation is linked to shifting cultivation[.]" But research in Sarawak by S.C. Chin, a Malaysian botanist who spent over ten years studying shifting cultivation, has shown that the typical family only cuts about two hectares per year, and that indigenous peoples as a whole are responsible for clearing about 72,000 hectares each year. It is estimated that only five per cent of this is virgin forest. Logging companies log around 450,000 hectares of primary forest annually.[5] Mitsubishi claims that by felling only four to six trees per hectare, they are practicing environmentally sound, selective logging by which the forest remains intact. Yet Chin estimates that even with this kind of operation, 40 per cent of the trees in each hectare are destroyed.[6]

Mitsubishi claims that forestry is essential to Malaysia's development aspirations – employing 55,000 people and earning hard currency. But benefits from the sale of timber concessions are dwarfed by the fact that 90 per cent of timber revenues are earned in the importing country, and most indigenous forest dwellers do not work directly in logging.[7]

Mitsubishi downplays its role in tropical timber trade, claiming that it does not rank among the top ten Japanese importers of tropical timber.[8] But if one includes figures for Meiwa Trading Co., which is controlled by Mitsubishi Keiretsu and Mitsubishi Corporation, and calculates imports of sawnwood and plywood as well as logs in roundwood equivalents, Mitsubishi/Meiwa was Japan's second largest importer of tropical timber in 1990.[9] In turn, Japan is the world's biggest importer of tropical timber, 92 per cent of which comes from Malaysia.[10]

"Much wasteful log use around the world is the result of local people's need to survive – and we cannot blame them for this."
– *Kyosuke Mori, Mitsubishi Corporation, 1991*[11]

"Our situation is like a child who has fallen into a fast flowing river and cannot swim. The child cries out, extending its arm for someone to help. If no takes the hand, the child will surely drown."
– *A Penan, on the environmental and*
  social crisis in Sarawak, 1991[12]

Mitsubishi's Reforestation Project: Irrelevant and Dubious

Since 1991, Mitsubishi has made much of its 50-hectare forest rehabilitation project in Bintulu, Sarawak. Mitsubishi touts the effort as "the first major scientific attempt to test techniques for restoring open barren areas to their natural state."[13]

Even if Mitsubishi succeeds in re-establishing some portion of the forest ecosystem in this 50 hectare area, the relevance of this example to endangered tropical forests across the globe is highly questionable. Given the significant costs involved, reforestation is hardly ever an economic option for less-industrialized countries. The costs per hectare of the Bintulu project are US$40,000 in the first three years alone. According to Malaysia's Federal Minister of Primary Industries, rehabilitation of all deforested areas in Sarawak would require an investment of US$60 billion. It should be noted that Mitsubishi has never committed itself to reforest Sarawak.[14]

Referring to the Bintulu project, Mitsubishi claims:"It is quite possible that a 'primeval' forest could exist on that land in 50-100 years."[15] The project uses only up to 25 tree species per hectare, however, whereas primary forests usually contain 200-250. It is extremely doubtful that Mitsubishi will be able to restore the vast and complex diversity of plant and animal life which logging has driven to extinction. Moreover, even if successful, the project would not address the severe adverse effects of deforestation on indigenous peoples.[16]

Mitsubishi and Logging Outside Malaysia

Malaysia's forests are not the only ones at risk from Mitsubishi's timber activities. Other areas of operation include:

*   Brazil, where Mitsubishi has a 49.5 per cent interest in Eido Do Brazil Madeiras S.A., a company which produces plywood and hardwood paneling, over half of which is exported to the US, Europe, and countries in the Caribbean. The company's operations extend into the Amazon as far as Benjamin Constant on the Peruvian and Colombian borders.[17]

*   Chile, where Mitsubishi joint venture Forestal Tierra Chilena Ltda. is growing eucalyptus for export to pulp mills in Japan.[18]

*   Bolivia, where Mitsubishi's joint venture Industria Maderera Suto Ltda. produces high-quality sliced veneer for export to Japan.[19]

*   Canada, where Mitsubishi owns a 30 per cent interest in Crestbrook Forest Industries Ltd., which through Alberta Pacific is clear-cutting Alberta's boreal forests, affecting many Dene and Cree people.[20]

*   Indonesia, where Mitsubishi is a major buyer of plywood. Mitsubishi markets Barito-Pacific Group's plywood in Japan through a network of over 400 factories and retailers.[21]

*   Papua New Guinea, where Mitsubishi funds United Timbers, which a commission of inquiry investigated and accused of hiding profits to avoid royalty payments.[22]

*   Siberia, where Mitsubishi is one of ten Japanese companies involved in a US$1.4 billion timber project. In addition, Mitsubishi has formed a joint venture with Hyundai of Korea called M&H Corp. which is importing silver fir into Japan and potentially endangering the last 200 Siberian tigers.[23]

Endnotes

1. Salomon Brothers, "Impact of Proposed BIS Rule Revision on Japanese Banks," November 1991.
2. L. Talbot, "A Proposal for the World Bank's Policy and Strategy for Tropical Moist Forests in Africa," 1990, quoted in Eric Wakker, Friends of the Earth Netherlands (FOEN), "Mitsubishi's Unsustainable Timber Trade: Sarawak," 1991, p. 5.
3. Rainforest Action Network (RAN), "An Examination of Mitsubishi's Involvement in Rainforest Destruction and the Impact of Environmentalists' Campaign Against Them," in *Boycott Mitsubishi Campaign: International Organizer's Manual*, San Francisco, CA, 1992, p. 12.
4. Wakker, op cit, pp. 4 & 12, and RAN, ibid, p. 8.
5. Fred Pearce, "Hit and Run in Sarawak," in the *New Scientist,* 12 May 1990.
6. Wakker, op cit. See also Pearce, op cit.
7. Wakker, op cit, p. 10-11.
8. Stephan Schmidheiny with the Business Council for Sustainable Development, *Changing Course: A Global Business Perspective on Development and the Environment,* The MIT Press, Cambridge, MA, 1992, p. 242.
9. Japan Tropical Forest Action Network/Friends of the Earch, Japan (JATAN), "Japan's Imports of Tropical Timber," (internal report), 1991, Tokyo. See also Japan Company Handbook, Toyo Keizai Inc., 1990.
10. Schmidheiny et al, op cit, p. 242.
11. From 19 June 1991 letter of Kyosuke Mori, General Manager of Mitsubishi's Environmental Affairs Department, to the European Rainforest Movement.
12. Quoted in Wakker, op cit, p. 9.
13. Wakker, op cit, p. 1.
14. RAN, op cit, pp. 10-11.
15. Schmidheiny et al, op cit, p. 244.
16. RAN, op cit, pp. 10-11.
17. RAN, op cit, p. 8.
18. See JATAN (Japan Tropical Forest Action Network/Friends of the Earth Japan), "Trading Company Profiles (internal report)," 1990, Tokyo, and JATAN, "Japan's Imports of Tropical Timber (internal report)," 1991, Tokyo.
19. RAN, op cit, p. 9.
20. Ibid, pp. 8-9, Greenpeace interview with Crestbrook, and Russel Mokhiber et al, "The 10 Worst Corporations of 1992," in *Multinational Monitor,* December 1992, p. 14.

21. RAN, op cit, p. 9.
22. Royal Commission of Inquiry into the Papua New Guinean Timber Industry, and RAN, op cit, p. 9.
23. RAN, op cit, p. 9.

Preserving nature is a good deal.

Aracruz and nature work together. The company plants 20 million eucalyptus trees per year, providing renewable resources to supply its pulp mill. Alongside the planted forests, 25,000 hectares of natural reserves, enriched with a million and a half native and fruit trees, grow in perfect harmony.

At Aracruz, environmental protection, as well as energy conservation, product safety and maximum productivity are major commitments to produce its high quality pulp.

This partnership with nature is one of the reasons why Aracruz will be able to double its pulp production to over 1 million tons by 1991.

A totally integrated company, from its own planted forests to its specialized port terminal, Aracruz is the world's leading producer of bleached eucalyptus pulp.

ARACRUZ CELULOSE S.A.
Highest Export Quality

Rua Lauro Müller, 116 - 40 · 22290 · Rio de Janeiro · Brazil · Telephone: (021) 541-6637 · Telex: (021) 22256 ACEL BR · Telefax: (021) 295-7943 · **Aracruz International Ltd.** · New Malden House, 4th Floor, 1 Blagdon Road, New Malden, Surrey, · KT3 4TB · England · Telephone: 01-336-2800 · Facsimile: 01-336-1998/2001 · Telex: 933974 ARINT G

Circle No. 427 on Reader Service Card

Source: "2000 and Beyond", The Pulp and Paper Industry, a Special Cooperative Editorial Feature from *Tappi Journal* and *Pulp and Paper Journal*, February 1990.

Greenwash Snapshot #15

ARACRUZ CELULOSE S.A.

A case study in deforestation, chlorinated emissions, and paper over-consumption.

Aracruz Celulose S.A. (Aracruz)
Chairman: Erling Lorentzen
Headquarters: Rua Lauro Muller, 116-40 22290 Rio de Janeiro, RJ Brazil
Tel: 55-21-541-6637 Fax: 55-21-295-7943
Main business: bleached eucalyptus pulp.

Aracruz is a member of the WBCSD and a signer of the ICC Rotterdam Charter.

"Preserving nature is a good deal."
– *from Aracruz advertisement* [1]

Brazil-based Aracruz is the world's largest exporter of bleached eucalyptus pulp. The company plants some 20 million eucalyptus tree a year. The caption above, set inside a photograph of planted forests alongside natural reserves, heads an advertisement in which Aracruz boasts of its ability to combine environmental protection with maximum productivity.

Aracruz has been singled out by several sources to represent business's potential for sustainable development. The Business Council for Sustainable Development (BCSD) wrote of Aracruz's activities as a case study in sustainable development. *The Financial Times of London* has called the company's work a "showcase for how a thorny environmental area can be turned into an economically feasible and environmentally sustainable development project."[2] And the International Chamber of

Commerce (ICC) included a "corporate environmental profile" of Aracruz in its book *From Ideas to Action* – which the company itself authored.[3]

Since the BCSD and others declared that Aracruz represents their approach to sustainable development, the company deserves scrutiny even though it is much newer and smaller than the other corporations discussed in this book. One cannot compare Aracruz's operations with the destruction wrought by the chemical giants. But there is more than one side to the Aracruz operation, and a close look shows much of the positive rhetoric about the company to be greenwash.

As presented by the Pulp and Paper Institute, the *Financial Times*, and in BCSD and ICC materials, the claims to environment and development sainthood for Aracruz can be summarized as follows:

Claim #1: *Aracruz is preventing deforestation*

The Aracruz complex is in the coastal state of Espirito Santo, which was once almost entirely covered with Atlantic rainforest. The main causes of the 90 per cent deforestation of the area were logging, coffee plantations, and cattle ranching. But Aracruz's record on forests is far from perfect. Analysis of an environmental impact report presented by Aracruz to expand its pulp production shows that at least 30 per cent of the municipality of Aracruz (around 20,000 hectares) had naturally generated secondary forest which was replaced by eucalyptus plantations. The plantations have halted regeneration of a multi-species ecosystem in degraded areas.[4]

Aracruz is not directly involved in cutting primary forests, and argues that by "giving this land [in Espirito Santo] economic function we are preventing further incursion into the Amazon."[5] The company does not currently operate in the Amazon, but it has considered expanding its activities to the Amazon area, with a large pulp mill in Maranhao and involvement in the Vale do Rio Doce Forest Centers project in Carajas, which intends to establish one million hectares of planted forests in the eastern Amazon.[6] Thus, promoting the idea that land must be used for "economic function," and thereby forcing subsistence and local communities into marginal areas, will in fact hasten the destruction of rainforests.

Claim #2: *Aracruz's pulp and paper production does not harm the environment*

Aracruz uses a 5-stage bleaching process which uses both elemental chlorine and chlorine doxide. The use of chlorine chemicals for bleaching in the pulp and paper industry is a major cause of organochlorine contamination. Aracruz plans to switch to the use of chlorine dioxide only, aided by a new oxygen bleaching system.[7] Complete substitution of elemental chlorine by chlorine dioxide will reduce but not eliminate organochlorine discharge (see IP case study).

The company was fined in March 1991 by the state environmental agency for toxic gas emissions. In September 1991, it was fined again for dumping untreated effluent into a stream through a discharge pipe which was not listed on the company's license application to the environmental agency.[8] Aracruz has admitted the existence of the discharge but denies that it is dangerous.

Claim #3: *Eucalyptus plantations are part of sustainable forest management*

Tree plantations are not forests; they are timber farms designed to produce a wood crop in the shortest possible time, regardless of social or ecological cost. The average age of trees harvested by Aracruz is seven years. The destructive impact of eucalyptus plantations on tropical soils, water tables, biodiversity, and thus local livelihoods and communities, is well documented. In the case of the Aracruz plantations, local communities are still struggling to have the social and economic costs of the scheme recognized and compensated.

Claim #4: *"Sustainably produced" paper is better than recycled paper*

The *Financial Times* quotes an Aracruz executive as saying that paper produced from "sustainable" forestry is better than recycled paper because "with recycled paper one does not know where the original came from. We need to break this myth that saving paper saves rainforests."[9]

It is true that recycled fibers have many sources, but saving paper in the North – both by recycling and using less – will save tropical and temperate rainforests. Annual paper consumption per person in the

United States is more than ten times that in Brazil.

Aracruz exports 80 per cent of its product to the North, where huge over-consumption of paper products is the driving force behind massive plantation and pulp-producing operations like that in Espirito Santo, and is also a cause of rainforest cutting.

Claim #5: *Aracruz is helping a region with economic stagnation and is contributing to social progress*

Aracruz has been the beneficiary of some dubious social and economic conditions. It started the Espirito Santo eucalyptus plantation in 1967 under a military dictatorship from which it received incentives for the establishment of tree farms, a pulp plant, and other infrastructure. The company plantations include around 15,000 hectares of Tupiniquim Indian lands to which the company gained the rights simply by asserting that the people living in the area were not Indians anymore.[10]

When an agreement between Aracruz and the state prohibited the company from buying more land to expand its plantations, Aracruz turned to promoting the establishment of eucalyptus plantations on land that was used for food production by participating in an extension program which was intended to support farmers, using the justification that they were giving an alternative to peasants.[11]

Court Action Against Aracruz

In an embarrassing blow to Aracruz's "green" image, a Brazilian federal judge ordered the company in November 1993 to suspend reforestation activities such as planting with eucalyptus. The judge's ruling was the result of an action brought against Aracruz and two government environmental protection agencies by the public prosecutor's office, which accused the company of damaging the Atlantic coastal forest. Recent satellite data shows that the Atlantic forest has lost an area equivalent to 13 soccer fields every hour since 1975. Aracruz appealed the ruling.[12]

"The example of Aracruz proves that enlightened environmental stewardship can be combined with corporate profitability."
– *Aracruz's "Corporate Environmental Profile,"*
  in the ICC's From Ideas to Action [13]

Endnotes

1. From advertisement in The Pulp and Paper Industry "2000 and Beyond," February 1990.
2. Christina Lamb, "Chopping Down Rainforest Myths," *The Financial Times of London,* 8 January 1992.
3. International Environmental Bureau of the ICC, *From Ideas to Action Business and Sustainable Development, The ICC Report on the Greening of Enterprise 92,* Special Edition for the UN Conference on Environment and Development, Oslo, Norway, 1992, pp. 258-259.
4. "Analyses of the Environmental Impact Report, Universidade Federal do Espirito Santo, Technological Institute.
5. Lamb, op cit.
6. Janice Menezes, "Aracruz Prepares Mega-Project," *Jornal do Brasil,* Rio de Janeiro, 25 May 1991.
7. Jim Young, "Market Pulp: A Giant Star is Born," in *Pulp and Paper International,* October 1991, p. 51.
8. "SEAMA Can Close Aracruz Celulose," *A Gazeta,* Vitoria, Espirito Santo, 18 September 1991. See also, Fernando Dantes "Aracruz Shows Its Formula for Sustainable Development," *Gazeta Mercantil,* Rio de Janeiro, 26 April 1991.
9. Lamb, op cit.
10. Rogerio Medeiros, *Espirito Santo: Maldicao Ecologica,* Rio de Janeiro, ASB Editora, 1983. See also video of research project of Federation of Social and Educational Assistance (FASE) entitled "Forest Exploitation in the North of Espirito Santo State and South of Bahia: Its Impacts and Alternatives on the People."
11. Claudio Rocha, "Espirito Santo May Have Eucalyptus Monoculture," *A Gazeta,* 22 March 1992,
12. Jan Rocha, "Brazilian Court Halts 'Green' Firm's Forestry," *The Guardian,* 26 November 1993.
13. *From Ideas to Action,* op cit, p. 259.

Everything you know about recycling, recycled.

James Walizer, 6, son of Jim Walizer, International Paper Recycling Supervisor.

Source: *The New Yorker*, 23 October 1995.

Greenwash Snapshot #16

INTERNATIONAL PAPER COMPANY

A case study in dioxin pollution and clean technology suppression.

International Paper Company (IP)
Chairman & Chief Executive Officer: John T. Dillon
Headquarters: Two Manhattanville Road Purchase, New York 10577, USA
Tel: 914-397-1500   Fax: 914-397-1596
Major businesses: pulp & paper; paperboard; wood products & timber.

International Paper has operations in 25 countries and markets its products in 130 nations. The company is the largest private landholder in the US, where it controls 6.3 million acres of timberland. It is a member of the WBCSD.

"[W]e believe environmental concern and actions are full compatible with being a profitable manufacturer of high-quality paper, wood, and related products."
— *IP's 1991 Annual Report*[1]

"International Paper is a company obsessed with short-term profits. Top management will let nothing – not fairness, not the workers, not the health of the community, not the air we breathe, not even the long-term interests of the company itself – stand in the way of next week's extra dollar."
— *from A Citizen's Report on International Paper Company, 1991*[2]

"Our experience with IP has been that the company has the most
negligent record in the paper industry, not only environmentally, but
also with regard to protecting the health and safety of its workforce. In
IP mill sites across the country, IP has been cited time and again by
federal and state agencies for both safety hazards at work and public
health hazards in the community."

–   *Frank Bragg, United Paperworkers International Union, 1991*[3]

International Paper, world's largest integrated paper manufacturer,
wants people to believe that it provides stewardship of the land while
seeking annual sales gains of eight to ten per cent. In fact, IP has been one
of the worst polluters in the pulp and paper business, an industry which
has poisoned the environment with many billions of pounds of extraor-
dinarily toxic chlorinated compounds including dioxins and has destroyed
forest ecosystems through short-sighted logging practices.

IP Responses To Dioxin Contamination

IP claims both that its "processes and products are and will continue to be
safe" but that a "prime objective" of the company is "lessening the
environmental impact of its processes and products."[4] IP's approach to
dioxin illustrates this contradictory attitude.

Dioxins are the most infamous of the many toxic chemicals discharged
by chlorine-based bleaching processes of conventional pulp and paper
mills. Dioxin belongs to the organochlorine group of chemicals, virtually
all of which are toxic, persistent, and likely to bioaccumulate. The effects
of dioxins include cancer, birth defects, and damage to the reproductive,
neurological, and digestive systems at the lowest doses ever tested.[5]

Until at least 1989, International Paper's 11 chlorine-based bleaching mills discharged annually an estimated 50 to 75 million pounds of
organochlorines into US waterways. A 1990 US Environmental Protection
Agency (EPA) study showed that four IP mills – more than any other
paper company – exposed people who ate small amounts of fish from
nearby rivers and streams to a lifetime cancer risk of one in 10,000 or
greater. IP's Georgetown, South Carolina mill posed the most severe
health threat, with of a risk of one in 50.[6]

Documents from the US EPA made public in 1987 confirmed the presence of dioxins in pulp and paper mill effluent. EPA also revealed that dioxins were contaminating almost all chlorine-bleached paper items including diapers, napkins, and coffee filters.[7] Other studies have shown that milk products packaged in chlorine-bleached paper containers can absorb dioxins in alarming amounts, a finding that prompted New Zealand to abandon their use. IP has largely dismissed dioxin's human health threat; the company's 1991 Annual Report calls concern about dioxin "significantly overstated."[8] The dismissal of dioxin's human health threat is part of a major industry initiative to cover up the hazards of dioxins (see Monsanto case study). IP has its own reasons for downplaying dioxin problems; in 1990-91, over 7,000 individuals in Mississippi filed lawsuits totaling US$6 billion against Georgia-Pacific Corporation and IP for alleged harm and risk from dioxin.[9]

IP says that since the mid-1980s it has aggressively pursued ways to reduce its dioxin discharges. But when researchers announced in 1991 that slowing the bleaching process could significantly cut discharges of dioxin and other chlorinated by-products, IP immediately said it "wasn't feasible" even though a competitor successfully implemented the process. "Apparently," said Frank Bragg of the United Paperworkers International Union (UPIU), "IP's drive for short-term profits won't allow it to slow production in the slightest degree, no matter what the environmental costs."[10]

IP and other pulp bleachers now claim to have cut dioxin releases substantially, a most welcome development if it can be verified. But dioxin is only one of approximately 1000 different organochlorines formed during the bleaching process, less than one-third of which have been assessed for their environmental toxicity. More of these deadly poisons will likely be identified in bleach plant waste and in household paper products.

IP has had the solution to these problems at hand for years. Chlorine-free bleaching technology exists and is being used by many mills in Europe. Patent documents show that the US paper industry was aware in the early 1970s of the environmental hazards of chlorine bleaching and had developed chlorine-free bleaching technologies.[11] Had IP and others invested in these technologies the US pulp and paper industry could have been chlorine-free in the 1980s, thus avoiding the discharge of billions of pounds of organochlorines.

IP first filed a chlorine-free patent application in 1978. In 1983, IP was awarded a patent for "chlorine free process for bleaching lignocellulosic pulp."[12] IP's patent application acknowledges that chlorine-based bleaching processes cause "serious pollution problems." It also claims that the resulting bleached pulp is strong enough for practical use and has brightness levels acceptable for a wide variety of uses. Asked to explain the reason for IP's failure to install the process at any of its mills, US EPA senior technical advisor Dan Bodein responded, "...I don't exactly know why."[13]

Forest Destruction

"I know of no nation and no people that have maintained on a sustainable basis plantation-managed trees beyond three rotations."
– Chris Maser, forest ecologist, Sierra Club, 1991[14]

Loggers cut an estimated 100 billion trees each year to supply wood to make paper products. To satisfy this demand, the pulp and paper industry has replaced natural forests with fast-growing monocultures such as pine. The creation of these timber farms has led to species extinction, soil impoverishment, an increase in pest infestation and disease, and reduced biodiversity of both plant and wildlife (see Aracruz case study).[15]

IP controls three million acres of pine plantations in the southern United States alone, plus interests in New Zealand, Chile, and other countries. Although IP writes of "managed care of its timberlands," the company does not have a sustainable yield policy.[16]

Emphasis on tree farms comes at the expense of conservation and recycling, both essential to help preserve the world's natural forests. Conservation is severely lacking in the US, for example, where excessive packaging is routine and the average citizen uses ten times the paper as the average Brazilian. This overconsumption of paper products in industrialized countries contributes some 40 per cent of all the garbage that is landfilled or incinerated, an enormous part of the waste disposal crisis.

Recycling generally gets more attention, and its advantages over virgin pulp are considerable: 74 per cent less air and 35 per cent less water pollution, one-half the energy consumption, 24 fewer trees per ton of paper, and less waste to landfills and incinerators.[17] Despite recycling's advantages, however, in 1991 IP bought only 80,000 tons of waste paper to be converted into recycled paper, a mere one per cent of the 7.3 million tons of paper IP's mills are designed to make annually.[18]

Pollution & Environmental Violations

Besides the environmental hazards inherent in IP's forest "management" and chlorine-bleaching, the company has been a well-known polluter. In 1988 and 1989, IP ranked second in the US pulp and paper industry for overall releases of toxic chemicals including methanol, toluene, acetone, dichloromethane, and chloroform. IP also ranks second in the industry for the number of Superfund sites – 14 – where it is a potentially responsible party. In 1991, IP had 20 facility-quarters out of compliance with air permits.[19]

In 1989, IP agreed to pay the US and the state of Maine US$990,000 for violations of the Clean Air Act and to install US$4.2 million worth of air pollution control equipment at its Jay, Maine, paper mill. Two years later, however, IP paid another US$885,000 to Maine after the state accused the same mill of unlawfully releasing pollutants into the land, air, and water.[20]

Also in 1991, IP agreed to pay a US$2.2 million criminal fine after pleading guilty to five felony counts at its Maine mill. Three of the felony counts involved hazardous waste violations under the Resource Conservation and Recovery Act (RCRA), while the other two charged that IP knowingly made false statements to state and federal authorities regarding company's hazardous waste disposal practices. The fine was the largest ever assessed in Maine for environmental law violations and the third biggest criminal fine for RCRA violations.[21]

Threats to Health & Safety

Between 1986 and 1991, IP paid US$1.3 million in fines to the US Department of Labor's Occupational Health and Safety Administration

(OSHA) for 1,329 violations. These fines were far higher than industry competitors. IP had an OSHA violation rate of 82.9 per 1,000 employees, the highest rate of the country's 50 largest companies.[22]

During 1987-88, IP management and workers were involved in a bitter 16 month strike at the Jay, Maine mill. When paperworkers refused to accept contracts requiring deep cuts in wages and health care – during a period of record company profits and sales – management locked them out of their jobs, hired inexperienced replacements, and eventually fired over 2,300 workers. During this period, health and safety conditions at the plant deteriorated.[23] Two toxic leaks in the first half of 1988 injured 15 workers and a third forced the evacuation of 4,000 area residents. In July 1988, the US Department of Labor fined IP US$873,220 for over 200 alleged health and safety violations. This was the largest such fine ever levied in the pulp and paper industry.[24]

In 1991, OSHA proposed US$803,000 in fines against IP's Moss Point, Mississippi, paper mill for 96 alleged violations. IP challenged the fines.[25]

Endnotes

1. International Paper Annual Report for 1991, p. 18.
2. *A Citizen's Report on International Paper Company*, presented to IP stockholders in May 1991.
3. From 29 April 1991 press release "UPIU and Greenpeace Protest Toxic Discharge."
4. First quote from 27 March 1990 letter to Ron Picard from Robert Bakarat, IP Product Manager; second quote from Annual Report, p. 17.
5. For more information on organochlorines see Joe Thornton, *The Product is the Poison–The Case for a Chlorine Phase-Out,* a Greenpeace USA Report, Washington, DC, 1991.
6. See George Lobsenz, "High dioxin risks cited at 20 paper mills," United Press International, 25 September 1990; "List of 20 paper mills with high dioxin risk," UPI 25 September 1990; and Lawrence Knutson article, Associated Press, 24 September 1990.
7. "EPA Links Dioxin to Paper Mills," in the *Christian Science Monitor,* 1 September 1987.

8. Annual Report, p. 36.

9. See Annual Report, p. 36; Glenn Ruffenach, "Georgia Pacific and International Paper Named in $2 Billion Pollution Lawsuit," Dow Jones News Service, 8 December 1990; Paul Kemezis, "Jury's Dioxin Decision Sparks Flood of Coypcat Suits Against Paper Mills," *Environment Week,* 14 February 1991.

10. From 29 April 1991 press release " UPIU and Greenpeace Protest Toxic Discharge." See also "Scientists Find Paper Mills Can Reduce Pollutants," *The Wall Street Journal,* 1 April 1991.

11. Greenpeace press release "Chlorine-Free Paper: Old Patents Reveal Paper Companies Knew All Along," 26 September 1991.

12. United States Patent, appl. no. 233,668, 11 February 1981, with abstract.

13. Quoted in *The Paperworker,* publication of UPIU, October 1991.

14. Sy Montgomery, "She sees the forest and trees," in *The Boston Globe,* 6 July 1992.

15. Ibid. See also Renate Kroesa, Greenpeace International, "Sustainable Paper Production: Specific Actions for Environmental Compatibility," May 1991; and John Ryan, "Conserving Biological Diversity," in *State of the World 1992,* Worldwatch Institute, Washington DC, 1992.

16. For quote see Annual Report, p. 18. Also *International Paper,* The Council on Economic Priorities Corporate Environmental Data Clearinghouse (CEP), New York, May 1992, p. 1.

17. Greenpeace International, *The Greenpeace Guide to Paper,* by Renate Kroesa, 1990, p. 40, and *The Medium is the Message,* by Mark Floegel, 1994, p. 19.

18. CEP, op cit, pp. 10 & 19.

19. CEP, op cit, pp. 2-3, 5, 7, 23.

20. "International Paper, US Reach Accord on Pollution," *The Wall Street Journal,* 2 August 1989; "International Paper Agrees to Pay Maine $885,000 to Settle Lawsuit Filed by State," *BNA Environment Daily,* 29 April 1991.

21. "International Paper Pleads Guilty, Will Pay $2.2 million for Waste Crimes," *BNA Environment Daily,* 19 July 1991.

22. CEP, op cit, pp. 2, 6, 20, 24, 35-36.

23. *A Citizen's Report,* op cit.

24. Information from Cathy Hinds, National Toxics Campaign Fund; and CEP, op cit, pp. 35-36.

25. CEP, op cit, p. 2.

Electric energy is a driver of economic progress. The demand for electric power continues to grow in less developed as well as industrialized nations to fuel growth and meet social needs.

Two ABB 600 megawatt
turbine generators at the
coal-fired Shidongkou
power plant on the Yangzte
River in China represent an
important step in improving
power technology in that
nation, where demand for
power is increasing.

Source: ABB 1991 Annual Report.

Greenwash Snapshot #17

ASEA BROWN BOVERI LTD

A case study in global warming, dumping in Central & Eastern Europe, and nuclear greenwash.

Asea Brown Boveri Ltd. (ABB)

Co-Chairmen: Peter Wallenberg & David de Pury

Headquarters: PO Box 8131, CH-8050 Zurich, Switzerland

Tel: 41-1-317-71-11 Fax: 41-1-317-73-21

Major businesses: electric power plant construction; power transmission design and manufacture; nuclear reactors.

ABB was formed by the 1988 merger between ASEA of Sweden and BBC Brown Boveri of Switzerland, and operates through 1,300 companies worldwide in over 140 countries. ABB is a member of the WBCSD and a signer of the ICC Rotterdam Charter.

ABB is the world's foremost electrical engineering company, the leader in its core businesses of electric power production and transmission equipment. The company claims to guide "global environmental policy" and boasts that it was a "major participant" at UNCED.[1] Swiss billionaire Stephan Schmidheiny, who sits on ABB's Board of Directors and is one of the company's largest private shareholders, also served as chief advisor on business and industry to the UNCED Secretariat.[2]

ABB suggests that western investment in Central and Eastern Europe will provide an impetus for more stringent pollution regulation.

But, from November 1991 to February 1992 the Polish government fined ABB heavily for dumping toxic wastes from its turbine factory in Elblag, Poland, on local farmland. This factory also discharged emissions two to three times the Polish norm (and 20 times the amount found in discharges from Swedish factories).

ABB says it is a "clean technology company" and a "world leader in responding to environmental protection needs."

But, it is marketing energy systems such as nuclear power and a new coal technology which emits nearly as much carbon dioxide as traditional coal-fired plants.

ABB in Poland

In recent years, lured by cheap labor, low production costs, and market potential, ABB has moved aggressively into Central and Eastern European countries and the former Soviet Union, buying up bankrupt factories and entering into joint ventures. According to the company executives, this is part of a long-term strategy to shift ABB's manufacturing and export base from Western to Central and Eastern Europe. ABB already has more workers in Poland than in France, Spain, or Denmark. At the end of 1991, ABB employed at least 15,000 people (seven per cent of its workforce) in the region and plans to increase that figure to 35,000 by the mid-1990s.[3]

ABB's Chief Executive Officer Percy Barnevik has spoken of "ecological disaster" in Central and Eastern Europe and the former Soviet Union as if this experience will help raise environmental awareness in ABB and other Western companies and provide an impetus for more strict pollution regulation. After less than two years, however, ABB had already established itself as a major polluter in one Polish city, Elblag.

In mid-1990, ABB became the majority partner in a joint venture with Zamech, previously a state-owned enterprise. The new company, ABB-Zamech, runs a 3500-employee turbine factory in Elblag. ABB agreed at the outset to build a landfill for the over 50,000 tons of waste the plant generates annually. This waste includes large amounts of hazardous material such as used molding sand laden with formaldehydes, cyanides, phenol, and synthetic resins. The company also agreed to install air pollution control equipment.

By the end of 1991, however, Polish officials were angered by ABB's failure to make good on its promises. ABB had yet to build the landfill and the Polish press reported that the company had dumped 12-15 tons of waste next to the factory, covered only with a plastic sheet. Experts claimed that poisonous leachate was contaminating groundwater. "If ABB-Zamech does not build a proper landfill for toxic wastes," Elblag's mayor asserted, "it will have to take them to Sweden."[4] City leaders remain skeptical that the landfill will ever be built.

Furthermore, Polish environmental inspectors discovered another site, on land leased from a local farmer, where since April 1991 representatives of ABB-Zamech had dumped 7,000 tons of spent moulding sand. Analysis showed levels of phenol which officials believed were contaminating groundwater, a potentially serious threat because the intake for a drinking water supply is less than two miles from the dump site. Samples of this waste also revealed the presence of environmentally damaging polynuclear aromatic hydrocarbons and phthalates as well as high levels of manganese and chromium.[5]

This case presents a clear case of double standards. In Sweden, one of ABB's home countries, the Environmental Protection Agency (EPA) says that waste sand which may contain phenols "should not be left in areas which are protected water sources or other places where water might be polluted." The Swedish EPA also states: "Waste sand from cast or hardened phenolic resin-bound moulds and cores should be dumped at sites which have a permit to receive this type of waste."[6]

Although ABB claimed that the waste was harmless, Polish government officials prohibited ABB from using the site as a landfill and fined the company about US$200,000.[7] Payment was deferred by the local government upon acceptance of a rehabilitation plan.[8]

Finally, because ABB had also failed to provide the promised air pollution reduction equipment, measurements of ABB-Zamech's smokestack emissions revealed dust concentrations two to three times the Polish norm (and 20 times the amount found in smoke discharges from Swedish factories) as well as levels of sulfur dioxide that were 24 times the norm.[9] In response to criticism, ABB repeated its pledge to install dust collectors in the smokestacks responsible for the pollutants.

None of this, however, has stopped ABB from claiming that the Zamech facility has "been transformed into a center of excellence for the manufacture of gas and steam turbines."[10]

"Money has to be made, we aren't the Red Cross."
– *Eberhard von Koerber, ABB Executive Vice-President*
for Eastern European countries,1990[11]

"Clean, Green" Technology

"As a 'clean technology' company, ABB is well positioned to provide
the technical solutions and systems necessary for sustainable development."
– *from ABB's 1991 Annual Report* [12]

ABB says it provides "clean" power generation systems and focuses
research on "green" technologies which can help solve global warming
and other environmental problems. The company claims to be a "world
leader in responding to environmental protection needs."[13]

Given this, it would be reasonable to expect that ABB devote a
significant portion of its multi-billion dollar R&D budget to safe, renewable
energy alternatives such as solar and wind power. No mention of these
alternatives, however, appears in the company's 1991 Annual Report.
Instead, the Annual Report points to substantial increases in R&D
spending in fossil fuel-based technologies that contribute heavily to
carbon dioxide emissions and global warming.

Burning coal produces far more carbon emissions than does the
combustion of other fossil fuels. For an equal amount of net energy, coal
releases about 25 per cent more carbon dioxide – the chief greenhouse gas
– than oil and about 75 per cent more than natural gas. Between 1975 and
1990, rising world coal use accounted for more than half of the global
increase in fossil-fuel carbon emissions; by 1990, coal was responsible
for 42 per cent of those emissions, the same percentage as oil.[14] Coal
combustion also releases nitrogen and sulfur oxides which contribute to
acid rain, the source of widespread forest damage in Europe, North
America, and Asia.

Despite these well-known environmental hazards, ABB recently
began a potentially lucrative (revenues already exceeding US$500 mil-
lion in 1991) new generation of coal technology – pressurized fluidized
bed combustion (PFBC) plants. ABB calls these "clean" because PFBC

plants release an estimated 59 per cent less sulfur dioxide and 50 per cent fewer nitrogen oxides than conventional coal facilities.

However, PFBC emissions of carbon dioxide are only marginally less than traditional coal-fired plants.[15] Thus, while PFBC plants might partially alleviate regional acid rain problems, their ability to mitigate global warming is insignificant.

ABB believes that coal reserves exist "in excess of 200 years," a further reason the company promotes "clean" coal as "a very attractive fuel choice."[16] The possible environmental consequences of using coal – "clean" or otherwise – for another 200 years are not discussed by the company.

Natural Gas

One of ABB's biggest businesses is in gas turbine power plants. The company reports that it has significantly increased R&D spending in gas turbine designs, another fossil-fuel based technology the company calls "clean." Compared with coal, natural gas does offer some economic and environmental advantages. State-of-the-art gas turbines produce power more cheaply than coal-fired plants, and with up to 65 per cent less carbon dioxide emissions. These turbines can emit fewer nitrogen oxides and virtually no sulfur compounds.

However, natural gas becomes dirtier when viewed in the context of production and development. Natural gas's main constituent is methane, a greenhouse gas with 63 times the Global Warming Potential (GWP) of carbon dioxide over a 20-year period (and 21 times carbon dioxide's GWP over the span of a century). Fossil fuel production contributes about one-fifth of methane emissions into the atmosphere, with between 25-50 per cent coming from natural gas pipeline leakage (coal mining accounts for a similar amount).[17]

Besides the problem of fugitive methane emissions, pipelines intrude into ecosystems with the threat of air, water, and land contamination. Offshore natural gas exploration and development degrade the environment in ways comparable to oil extraction: drilling muds; "toxic brine"; air pollution; and wetland loss. Tanker transport of highly flammable gas is dangerous due to the risk of explosion and toxic release. Natural gas processing facilities, because they deal with many combustible and toxic materials, pose hazards to the environment as well as to public health.[18]

Both those who advocate an immediate phase-out of all fossil fuels and those who see a transitional role for natural gas in the switch to renewables will be disappointed to find that ABB's commitment to natural gas is not contingent upon renovation of pipelines to eliminate methane emissions, nor upon a "bridging" role for gas in the effort to develop renewable energy alternatives. In the end, ABB's coal and natural gas practices mimic the oil industry's stated intention to develop all the oil in the world.[19] ABB calls these practices, which threaten the stability of the earth's climate, "sustainable," and "clean."

"Ultra-safe" Nuclear Power?

ABB predicts a revival in the fortunes of nuclear energy and has built a new reactor called PIUS (Process Inherent Ultimate Safety) that incorporates "inherent" or "passive" safety systems which rely on natural processes rather than machinery to operate. ABB describes this technology as "ultra-safe."[20] However, the UK's Atomic Energy Authority (AEA) has conducted one of the few systematic evaluations of these designs and concluded that while simpler, "inherent" systems are not necessarily safer than traditional systems. Engineers designed "inherent" systems to prevent causes of past accidents, according to the AEA, but have ignored other known problems which could lead to future disasters. Moreover, conventional reactors and those with "inherent" systems are equally vulnerable to structural failures. [21]

ABB says that "environmental concerns" about carbon dioxide emissions "favor" nuclear power, the implication being that nuclear power can help stop global warming. This position is extremely misleading, and is one of the most dangerous pieces of greenwash yet concocted by industry (see Westinghouse case study).

ABB and Iraqi Nuclear Buildup

In December 1991, the International Atomic Energy Agency (IAEA) published a list of foreign supplies and suppliers to the Iraqi nuclear weapons program. Among the equipment found was a large cold isostatic press manufactured by ABB. The IAEA said that this press, like the other equipment, was "application-specific" and asserted further: "While much of the equipment is multi-purpose in the sense of being useful in a number of manufacturing processes, the presence of application specific features removes most doubt as to the intended use." [22]

ABB acknowledged that it sold the press to Iraq in 1990 but said the Iraqis had guaranteed that it would not be used to make nuclear weapons. "[I]t's unfortunate that the equipment was found where its use can be questioned," an ABB spokesperson explained.[23]

ABB and Hydro-Quebec

Although it receives short shrift in comparison with coal, gas, and nuclear power, "hydro-generation" also falls under ABB's umbrella of "sustainable" electricity sources. The company has provided equipment to a number of large hydro-power projects worldwide, although such projects are well-known for flooding forests, ruining wildlife habitats, and uprooting entire communities of indigenous peoples.[24]

Hydro-Quebec's James Bay scheme, for example, was the largest and potentially most destructive hydroelectric project in North American history. Cancelled in late 1994 due to intense opposition, the project had threatened to displace 18,000 Cree and Inuit people, destroy 11 rivers, and adversely affect a wilderness area the size of France. Despite these well-known consequences, ABB was willing to do business with Hydro-Quebec. In 1989, ABB Canada signed a joint venture with Hydro-Quebec to develop transmission technology. The next year, ABB received a US$100 million contract from Hydro-Quebec to install electrical equipment. And in 1991, ABB's subsidiaries in Canada and Sweden signed a US$30 million contract with Hydro-Quebec to supply additional electrical equipment.[25]

Endnotes

1. ABB Annual Report 1991, p. 26.

2. According to the French journal *Bilan* ("Dossier: Stephan Schmidheiny," November 1990, p. 135), Schmidheiny owns an 18 per cent stake in BBC Brown Boveri Ltd. Based on BBC's 50 per cent ownership of ABB Asea Brown Boveri Ltd., Schmidheiny may own nine per cent of ABB.

3. See Kevin Liffey, "Eastern Europe: Zurich-based ABB seeks to electrify reforming Eastern Europe," Reuter News Service, 16 October 1990; and "Finland: ABB chief sees Baltic area as key economic region," Reuter News Service, 15 January 1992.

4. "Polish-Swedish Waste," in *Gazeta Wyborcza,* 5 December 1992.

5. Iza Kruszewska, "Open Borders, Broken Promises Privatization and Foreign Investment: Protecting the Environment Through Contractual Clauses," Greenpeace International, Amsterdam, 1993, pp. 38-40. Tests results revealing phenol content from WIOS (County Pollution Inspectorate), 11 November 1991.

6. Ibid, p. 41.

7. Ibid, p. 40. Also Iza Kruszewska, Greenpeace International, correspondence from December 1991.

8. Ibid, p. 40. Also Kruszewska correspondence, July 1992.

9. Ibid, p. 41. Also Kruszewska correspondence, December 1991, and article from Swedish magazine *SIF tidningen,* 3 May 1992.

10. From advertisement in *The Economist*, 27 March 1993.

11. Liffey, op cit.

12. ABB's Annual Report 1991, p. 27.

13. Ibid, p. 26.

14. Christopher Flavin, "Building a Bridge to Sustainable Energy," in *State of the World 1992*, a Worldwatch Institute Report, Washington, DC, 1992, p. 30. For other information on coal see Peter Ciborowski, "Sources, Sinks, Trends, and Opportunities," in *The Challenge of Global Warming,* D.E Abrahamson, ed., Washington, DC & Covelo, CA, 1989, pp. 213-230.

15. According to ABB (quoting from a 13 August 1991 article in *Energy Daily*), PFBC plants release ten per cent less CO_2 than conventional coal plants. This may be optimistic; according to Flavin, ibid, p. 36, PFBC plants emit *more* CO_2 than conventional ones. The Applied Energy Systems Corporation, Arlington, VA, has compiled a "Comparison of Carbon Emissions From Fossil Fuels" study. According to Jennifer Lowry of AES, PFBC plants fall within the parameters of carbon emissions listed for all coal

plants. Information on other PFBC emissions from Bob Travers of the Office of Fossil Energy of the Department of Energy.

16. ABB 1991 Annual Report, pp. 14 & 40.

17. *Environmental Data Report,* 2nd edition, United Nations Environment Programme, London, 1989-90, pp. 6-7. See also *The Ecologist,* vol. 21, no. 4, July/August 1991, p. 162, for more information on methane sources.

18. Carol Alexander, Greenpeace, *Natural Gas: Bridging Fuel? Or Road Block To Clean Energy?,* pp. 9-14.

19. This intention from Greenpeace personal interview with ENI CEO Cagliari and Shell representative, 1992. For impact of oil production and consumption see Greenwash Snapshot #1 on Shell.

20. ABB 1991 Annual Report, p. 15.

21. "Outlook on Advanced Reactors," in *Nucleonics Week,* 30 March 1989, pp. 7-10. See also Christopher Flavin, "Slowing Global Warming: A Worldwide Strategy," Worldwatch Paper 91, October 1989, pp. 36-37.

22. IAEA Press Release "Foreign Supplies To The Iraqi Nuclear Programme," 11 December 1991.

23. "Sweden: ABB says it sold equipment found in Iraqi Nuclear Programme," Reuter News Service, 12 December 1991.

24. See Cynthia Pollock Shea, "Renewable Energy: Today's Contribution, Tomorrow's Promise," Worldwatch Paper 81, January 1988, pp. 12-14. See also Steve Turner and Todd Nachowitz, "The Damming of Native Lands," in *The Nation,* 21 October 1991.

25. "Canada: ABB power systems of Sweden wins order from Hydro-Quebec," *Svenska Dagbladet,* 23 August 1991; "Canada: Asea Brown Boveri wins $30 million contract with Hydro-Quebec," *Financial Post,* 4 July 1991; *Canadian Chemical News,* April 1990; and "ABB To Agree to $310 Million as Price in Spinoff," *Wall Street Journal* (Europe Edition), 16 December 1989.

Trees aren't the only plants that are good for the atmosphere.

Because nuclear plants don't burn anything to make electricity, nuclear plants don't pollute the air.

In fact, America's 111 operating nuclear electric plants displace other power sources and so reduce certain airborne pollutants in the U.S. by more than 19,000 tons every day. Just as important, nuclear plants produce no greenhouse gases.

But more plants are needed—to help satisfy the nation's growing need for electricity without sacrificing the quality of our environment. For a free booklet on nuclear energy, write to the U.S. Council for Energy Awareness, P.O. Box 66080, Dept. HP01, Washington, D.C. 20035.

Nuclear energy means cleaner air.

© 1991 USCEA

As seen in January issues of Reader's Digest, Newsweek, Smithsonian, Forbes, Atlantic and Natural History; February issues of TIME, Christian Science Monitor, World Monitor, The Washington Post, The Washington Post National Weekly Edition and American Heritage; and March issues of Ladies' Home Journal, Good Housekeeping, FORTUNE, New Choices, Congressional Quarterly and National Journal.

Source: Advertisement published by the US Council for Energy Awareness, 1991.

WESTINGHOUSE ELECTRIC CORPORATION

A case study in nuclear power construction.

Westinghouse Electric Corporation (Westinghouse)
Chief Executive Officer: Michael H. Jordan
Headquarters: Westinghouse Building Gateway Center Pittsburgh,
Pennsylvania 15222, USA
Tel: 412-244-2000   Fax: 412-642-3404

Major businesses: nuclear weapons; electronics; electrical supplies; nuclear power; waste disposal.

Westinghouse operates in 36 countries.

In the US, when people hear the name "Westinghouse" they think of household appliances. Only rarely does the company publicize another side of its business: nuclear weapons and reactors. Westinghouse produces nuclear propulsion systems for military submarines and nuclear-armed surface ships as well as launching systems for intercontinental ballistic missiles and cruise missiles such as the MX and Trident.[1] The company has operated nuclear weapons facilities for the US government at: Hanford, Washington; Fernald, Ohio; Idaho Falls, Idaho; and Savannah River, South Carolina. The most optimistic estimate for cleanup costs for contamination at Hanford alone is US$30 billion over 30 years, if indeed cleanup can be done at all.[2]

The environmental damage at US nuclear weapons production facilities is so monumental that it cannot be greenwashed away. But in its civilian businesses, Westinghouse has joined the corporate greenwash brigade.

Nuclear Greenwash

"Our industry [nuclear power] will become one of the principal symbols of the entire environmental decade. We can be cast as its hero."
— *Richard Slember, Vice President & General Manager*
   Westinghouse's Energy Systems Business Unit, 1990[3]

A pioneer in the development of nuclear reactors throughout the world, Westinghouse now touts nuclear power not only as the answer to global energy needs but also as a savior of the environment. Because they burn no fossil fuels, Westinghouse suggests, nuclear plants are an answer to the greenhouse effect.[4] The US nuclear lobby which Westinghouse helps fund says that nuclear power can help stop global warming while it satisfies the demand for electricity. An advertisement by the nuclear lobby group US Council for Energy Awareness showed a bucolic country scene with a nuclear plant in the background and the caption: "Trees aren't the only plants that are good for the atmosphere."[5] This is greenwash at its most absurd.

In practical terms, nuclear power cannot stop global warming. Nuclear reactors generate electricity and only about one-sixth of greenhouse gases come from burning fossil fuels for electricity. Research has shown that a crash program to offset carbon emissions from coal-fired electricity generators alone would require the construction of 5,000 nuclear reactors over the next three decades, most of them in the South, at a cost of US$144 billion in capital expenses annually and electricity generation costs of US$525 billion per year.[6] Even with such a construction program, carbon dioxide emissions globally would still rise.

In fact, analysts have theorized that spending on nuclear power can worsen global warming by draining energy investment away from energy efficiency and renewables programs. For each US$100 spent on nuclear power, one metric ton of carbon is effectively released in to the atmosphere which could have been avoided had the money been spent on energy efficiency. Additionally, such a nuclear construction program would generate 100,000 tons of high-level radioactive waste per year, with no resolution to the nuclear waste crisis in sight.[7]

"If we can't resolve what we're going to do with the waste, then we have no business generating it."
  – *Cecil Andrus, former US Secretary of the Interior* [8]

Hazards of Nuclear Power

Besides radioactive waste, nuclear power poses other threats to people and the environment. The risk of exposure to radiation from routine emissions of contaminated gas and water from nuclear plants as well as from the transportation and storage of radioactive materials is always present. A 1990 study by the US National Academy of Sciences raised official estimates of the dangers of low-level radiation by 350 per cent. "Don't have children" was the official advice given to some British nuclear workers after a government study showed that their offspring were seven times more likely to develop cancer. [9]

The potential consequences of nuclear power plant accidents are devastating. Estimates of the number of deaths over the next 70 years from radiation from the Chernobyl disaster range from official predictions of 40,000 to independent ones of 280,000 to 500,000. Authorities admit that at least 100,000 people continue to live in contaminated areas. Water can no longer be used within 400 kilometers of the accident site, food is imported, and more than 50,000 square kilometers of arable land have been abandoned. According to recent findings by a team of scientists from the Ukrainian Academy of Sciences, the radioactive effects of the accident will continue for up to 24,000 years. A study in the former Soviet Union calculated the damage from Chernobyl to be US$283-358 billion; the total bill suggests that the country might have been better off if it had never built a single nuclear reactor. [10]

Westinghouse in the Philippines

Westinghouse and other companies have promoted and exported nuclear power technology to countries in Latin America and Asia.

The Philippines' experience with Westinghouse nuclear power plant construction is notorious and should serve as a warning to other countries targeted by the company for such investment. In 1976, Westinghouse signed a US$1.1 billion contract with President Ferdinand Marcos to build a nuclear reactor in the Bataan peninsula. By 1985, when Westinghouse completed construction, the Bataan plant's price had risen to US$2.2 billion, or over eight per cent of the Philippines's foreign debt. Interest payments alone on loans for the plant are about US$350,000 a day. The plant is the largest single capital investment in Philippine history – in a country where poverty, unemployment, and malnutrition are pervasive.[11]

Seventy per cent of the Philippine population, primarily landless tenants and subsistence farmers, consume a tiny fraction of the country's electricity. "The reactor is not designed to supply electricity to our people," a Philippine activist pointed out in 1979, "It's for Clark Air Force Base and the Subic Naval Base and the Bataan free trade zone, where foreign companies make textiles for foreign markets – most of them American."[12] Hardest hit by the Bataan reactor were the 13,000 poor villagers displaced by the plant's construction.

Marcos's successor, President Corazon Aquino, refused to operate the reactor, the safety of whose design and construction had come under question years earlier. In 1988, the Philippine government initiated a lawsuit against Westinghouse alleging that the company paid bribes to an associate of Marcos's in order to win the plant construction contract. That case was suspended in 1992 when Westinghouse proposed to pay US$100 million in cash and services to the Aquino government. Even with new investment of US$400 million in needed repairs being considered, the plant would not be operable for at least three years.[13]

This proposal immediately came under fire in the Philippine Congress. Lawmakers expressed doubts that the plant could be made safe and criticized the plan because it would force the Philippines to borrow an additional US$325 million from the US Export-Import Bank for the repair work. In December 1992, Aquino's successor President Fidel Ramos ordered prosecutors to resume litigation. Six months later, in May 1993, a US jury cleared Westinghouse of the bribery charges. "We should obtain vindication someday," President Ramos asserted, "Bribery is a crime committed in the darkness. The conspirators do not do it in the sunlight."[14]

The Philippine government appealed the verdict, and launched a

US$40 million damage suit against Westinghouse at the International Arbitration Court in Geneva, Switzerland. The government also announced that it would convert the Bataan plant into a coal-fired or combined-cycle facility. In late 1995, the Philippine government agreed to drop the 1988 lawsuit after reaching a settlement with Westinghouse. Terms of the settlement were not disclosed.[15]

Meanwhile, the Philippines continued to pay interest on the US$2.2 billion debt to Westinghouse, and the country has suffered from severe power shortages which have been blamed on the mothballing of the plant. At times, Manila has had about eight hours of power outages each day because of deficient generating capacity.[16]

> "Many of us have no more land to till. The lands where we used to get our food and livelihood from are either bought at low price or confiscated because they were needed by the plant. Before, the fishermen used to fish near the shore.....[The government] has driven the fish away because earth fillings are washed directly into the sea. Parts of mountains abundant in fruit trees and other crops are already leveled off."
> – *Bataan resident, about the effects of the nuclear plant construction, 1979*[17]

> "[A Marcos-appointed commission] concluded that Westinghouse had shown a lack of immediate concern over the safety of the plant."
> – *Robert Pollard, Nuclear Safety Engineer, Union of Concerned Scientists, citing the findings of the Puno Commission, 1981*[18]

> "We're proud of the plant we built."
> – *Nathaniel Woodson, General Manager Westinghouse's Nuclear Fuel Business, speaking about the Bataan reactor, 1988*[19]

Problems with Westinghouse Nuclear Plant Construction elsewhere

The construction problems with the Bataan plant are not an isolated case. Brazil's Furnas Centrais Electricas was engaged in a dispute with Westinghouse over alleged defects with the company's steam generator at the nuclear reactor Angra-1, Brazil's first such plant. Construction of the reactor was projected to be five years but took ten due to difficulties aggravated by poor initial management. The original projection for the plant's cost was US$829 million; the final cost came to almost US$3.5 billion, raising Brazil's foreign debt significantly.[20] Estimates of losses the utility has suffered because of interest and service payments are around US$500 million.[21]

Angra-1 was shut down in March 1993 because of fears about its safety. Two months later, technicians confirmed that defects in the nuclear fuel rods were producing leaks in the plant's water circuit, which had a radioactivity level far above normal.[22]

Westinghouse reactors in the US have also come under criticism. In 1993, Portland (Oregon, US) General Electric Company initiated a lawsuit against Westinghouse over allegedly faulty equipment which caused the permanent closure of the 17-year-old Trojan Nuclear Plant. The suit claimed the equipment contained "serious defects in design, material, and workmanship" and was "not suitable for operation." The suit also alleged that Westinghouse knew about problems with the equipment as early as 1968, when it was sold to the utility company.[23] The case is pending.

Moreover, 11 additional US utility companies have filed suit against Westinghouse for alleged problems with its nuclear reactors. While these cases are pending, Westinghouse settled yet another lawsuit by Carolina Power and Light Company in June 1993 for alleged defects in steam generators of two reactors. Terms of the settlement were not disclosed.[24]

Westinghouse in Central and Eastern Europe

Faced with stagnant demand in the industrialized world, Westinghouse is currently promoting nuclear power in Central and Eastern Europe. In 1993, Westinghouse said it would use its own technology to complete two reactors in Temelin, the Czech Republic. In March 1994, with the Clinton Administration's backing, the US Export-Import Bank approved a US$317 million commercial loan guarantee to finance the project.[25] Some US lawmakers raised the concern that, because of the Export-Import Bank's guarantees, taxpayers in the United States might be liable for damages were there an accident at the Temelin facility.[26]

The reactor model involved at Temelin is of Soviet design. The Export-Import Bank, when it approved the loan guarantee, said it used its own safety analysis as well as assessments by the US Nuclear Regulatory Commission (NRC) and the International Atomic Energy Agency (IAEA).[27] The NRC had not performed an evaluation of this type of reactor, however, while the IAEA did study it and found design problems but had not determined whether Westinghouse's technology could repair the difficulties.[28] Neither an environmental impact assessment nor an independent technical safety review of Westinghouse's plan to complete the reactors' construction with US equipment has been published.[29]

Local opposition to Temelin plant has been strong. Some 60 towns and villages in the Temelin area formed an association and took a poll on the issue; representatives from all but four villages expressed their solid opposition to the plant.[30] The Austrian government – Temelin is 120 miles from Vienna – has also registered great concerns about the project and the possibility of an accident.[31]

Despite such resistance, Westinghouse continues to seek targets in Central and Eastern Europe for exporting nuclear technology. It has bid on a contract for work on four Soviet-designed reactors in Hungary, and was reported to be a finalist in the bidding. This project would also involve Export-Import Bank financing.[32]

Endnotes

1. See Jonathan Pressler, "The Other Westinghouse: Weapons and Waste," The River City Nonviolent Resistance Campaign, Pittsburgh, 10 April 1989.
2. Thomas Lippman, "Facing a Nightmare of Poisoned Earth," *The Washington Post,* 2 December 1991.
3. For Slember's remarks made at the 1990 meeting of the Nuclear Power Assembly see 23 May 1990 PR Newswire. See alsoWestinghouse's "Westinghouse and the AP600 – Meeting the Challenge for Clean, Safe, Economic Power."
4. Ibid, "Westinghouse and the AP600."
5. US Council for Energy Awarness advertisement, 1991.
6. B. Keepin and Kats, G., "Greenhouse Warming: Comparative Analysis of

Nuclear and Efficiency Abatement Strategies," in *Energy Policy,* December 1988, vol. 16, no. 6, pp. 538-61.

7. According to Nicholas Lenssen in "Nuclear Waste: The Problem that Won't Go Away," paper 106, Worldwatch Institute, a nuclear reactor produces on average 20 tons of radioactive waste per year.

8. Quoted in "The Other Westinghouse," p. 13.

9. For this and other information about the potential hazards of nuclear power, see Greenpeace Action's Factsheet "Nuclear Power," Washington, DC, 1991.

10. For offical Soviet estimate see *International Herald Tribune,* 25 August 1988. For 280,000 estimate see Richard Webb, *The Ecologist,* v. 16, no. 4, 1986. For 500,000 figure see calculation of Professor John Gofman in *International Herald Tribune,* 11 September 1986. For other information see: *The Times,* 8 November 1989; V. Haynes and M. Bojcun, "El Desastre de Chernobyl," 1988; Greenpeace America Latina, "Un Peligro Imminente–Atucha-1," Buenos Aires, 1990, p. 5; and Richard Hudson, *The Wall Street Journal,* 29 March 1989. For findings by Ukrainian scientists see Andrei Ivanov and J. Perera, "Ukraine: Chernobyl will linger for 24,000 years," IPS, 24 June 1993. See also Greenpeace Action, "Nuclear Power" factsheet.

11. Brian Dumaine, "Managing the $2.2 Billion Nuclear Fiasco," in *Fortune,* 1 September 1986. See also Associated Press report "Philippines to Sue Westinghouse," 27 October 1988.

12. Quoted in Harvey Wasserman, "Radiation Roulette – Once You Start Gambling on Foreign Nuclear Reactors...How Do You Stop?" in *Mother Jones,* August 1979, p. 57.

13. Holley Knaus, "Westinghouse Settles," in *Multinational Monitor,* April 1992, p. 4.

14. See Michael Di Cicco, "Ramos wants to convert plant, plan may be unfeasible," UPI, 7 July 1992, and "Philippines Blasts Westinghouse Ruling," Reuters, 19 May 1993.

15. Ramon Isberto, "Philippines: Facing Anti-Nuke Mindsets," IPS, 2 June 1993, and "Manila not keen on deal with Westinghouse," Reuters, 22 November 1994. See also "Westinghouse Pact with Philippines," *The New York Times,* 16 October 1995.

16. Reuters, op cit.

17. Quoted in E. San Juan, Jr., "Blueprint for Disaster –Westinghouse Brings Nukes to the Phillipines," in *Science for the People,* January/February 1980, p. 24.

18. Quoted in Address of R. Pollard, Nuclear Safety Engineer, Union of Concerned Scientists, "On the Bataan Nuclear Power Plant," p. 4.

19. Quoted in Fox Butterfield, "Philippines Expected to File Suit Against Westinghouse," *The New York Times,* 1 December 1988.

20. Ruy de Goes, Greenpeace Brazil.

21. "Inside N.R.C.," 20 July 1987.

22. The British Broadcasting Corporation Summary of World Broadcasts, 18 May 1993, "Radioactive fuel leak in power plant," Globo TV, Rio de Janeiro, 8 May 1993.

23. "Portland General sues Westinghouse over nuclear plant," UPI, 17 February 1993, and William McCall, "Trojan-Westinghouse," AP, 17 February 1993.

24. Dave Airozo, "NSP Sues Westinghouse over Prairie Island Steam Generators," in *Nucleonics Week,* 5 August 1993.

25. Douglas Frantz, "US Backing Work on Czech Reactors by Westinghouse," in *The New York Times,* 22 May 1994.

26. Ibid.

27. Ibid.

28. Ibid.

29. Letter from the Temelin Task Force, a coalition including Greenpeace, Nuclear Information and Resource Service, the Safe Energy Communication Council, Friends of the Earth USA, Global 2000, Public Citizen, and the Institute for Policy Studies, to US President Clinton, 5 May 1993.

30. Ibid.

31. Frantz, op cit.

32. Frantz, op cit.

Source: From press packet on Heinz's "dolphin safe" policy.

Greenwash Snapshot #19

H. J. Heinz Company/StarKist

A case study in "dolphin safety."

H.J. Heinz Company (Heinz)
Chairman: Anthony J. F. O' Reilly
Headquarters: 600 Grant Street Pittsburgh, Pennsylvania 15219, USA
Tel: 412-456-5700 Fax: 412-237-5377
Major businesses: food and beverages; condiments and sauces; pet food.
Major subsidiary: StarKist Foods, Inc.

One of the world's largest food processing companies, Heinz's facilities span six continents; recent expansion plans include Eastern Europe and South Africa. Over 40 per cent of Heinz's sales come from non-US operations.

> "StarKist clearly recognizes the growing public concern for protecting dolphin lives, and the concern that current levels of dolphin mortality are too high. StarKist has therefore decided to take the leadership role in protecting dolphin lives."
> – *From "Common Questions Concerning 'Dolphin Safe' Tuna"*[1]

In April 1990, just before the 20th anniversary celebration of "Earth Day", Heinz proclaimed that its subsidiary StarKist would henceforth adhere to a "dolphin safe" policy. The company declared it would no longer buy tuna fish caught in association with dolphin in the Eastern Tropical Pacific Ocean. To remind the public of this commitment, labels on cans of tuna sold in the US would bear a "dolphin safe" symbol. Following Heinz/StarKist's lead, two other major canning companies issued similar policies.

Heinz/StarKist's well-timed announcement drew much attention in the US, where one-third of the world tuna catch was consumed and where StarKist controlled the largest share of the canned tuna market. National politicians, the media, and environmentalists applauded. At the time, the decision seemed to indicate that Heinz/StarKist was encouraging a conversion from destructive to environmentally sound fishing practices.

Heinz/StarKist's subsequent behavior has belied this initial promise and reveals the true nature of the "dolphin safe" policy: a marketing strategy designed less to safeguard dolphin than to protect the company's public image and economic interests. Because Heinz/StarKist is the world's biggest tuna canner, it has a huge stake in the success of that strategy, and has spent much time, energy, and money to promote itself as a savior of dolphin.

Meanwhile, Heinz/StarKist has invested nothing in international programs to develop sound alternative fishing technologies which would not threaten dolphin or other vulnerable marine life. Heinz/StarKist has opposed environmental regulation to enforce and verify the safety of dolphin in most of the world's ocean areas. In addition, Heinz/StarKist and other transnational tuna corporations have used the "dolphin safe" label to avoid dealing with the larger global impact of commercial tuna fishing.

Using Dolphin To Catch Tuna

The commercial fishing industry has employed essentially the same method to catch tuna fish since the late 1950s, when it was developed. That method is called "purse seining".

It is well known that tuna sometimes swim below schools of dolphin. As a result, dolphin are used to detect the presence of tuna. Once a dolphin school is spotted, speedboats chase and herd the group – "much like herding cattle," according to Heinz/StarKist literature – to concentrate the tuna.[2] Sometimes, underwater explosives are used to disorient and help corral the dolphin. The chase ends when the exhausted dolphin slow down and form a protective circle.

A large circular "purse seine" net, so called because it has a purse-like cable drawstring on the bottom, is cast around the dolphin (this is known as encirclement). The drawstring is tightened, the net closes, and the tuna – and dolphin – are entrapped.

Dolphin can become entangled in the netting and suffocate. They may be crushed by the weight of the tuna in the net, or by the power winch which hauls in the net. As dolphins struggle to free themselves, their flippers or beaks can be ripped off. Dead or dying dolphin are tossed back into the sea as "waste." Since 1959, the commercial tuna industry has been responsible for the death of over seven million dolphin. The US fishing fleet, a part of which was owned by StarKist for years, accounted for almost 89 per cent of all dolphin deaths that have occurred in one fishery called the Eastern Tropical Pacific (ETP) Ocean, a fishery in national and international waters stretching from Mexico to Chile, where the tuna-dolphin association is best known.[3]

Limits to "Dolphin Safe"

"StarKist's policy is simple. If nets are not set on dolphins, then dolphins will not be killed or injured."
–   *From "StarKist Dolphin Safe" brochure*[4]

I. *Geographic Limits*

Heinz/StarKist's policy applies to tuna caught in association with dolphin in one fishing area, the Eastern Tropical Pacific. Commercial tuna fishing in the ETP is more regulated – with observers on board every vessel in the fishery – than in any other tuna fishing area.

However, research shows that tuna and dolphin swim together in tropical fishing grounds around the world.[5] Ocean areas outside the ETP, which supply the vast majority of canned tuna and where the setting on dolphin is occurring, remain largely or completely unregulated.

In the Eastern Pacific, a precedent-setting international program passed in 1992 by the Inter-American Tropical Tuna Commission appears to be one solution to the extreme dolphin deaths in the fishery. Heinz/StarKist never supported calls to establish this international program for the conservation of dolphin, which Southern-based environmental organizations and Greenpeace advocated. The program, in two years of operation, has reduced dolphin mortality from around 100,000 animals a year to less than 4,000 in 1994. The multilateral regime regulates the tuna fishery, requires 100 per cent observer coverage for the international fleet, establishes an international panel of governments, environmentalists, and industry which oversees every fishing trip to the Eastern Pacific, and sets up an international research panel to investigate alternative fishing gears.[6]

StarKist has not merely failed to advocate this regulatory program in the Eastern Pacific. Additionally, it has lobbied consistently against observer coverage, monitoring, enforcement, or any other kind of regulation in tuna purse seine fisheries in the South Pacific, East and West Atlantic, and Indian Oceans.[7]

Heinz/StarKist has also encouraged vessels to migrate to fisheries where monitoring and control are, for the most part, not operating and where the environmental impact of commercial fishing is probably severe. StarKist has even sought governmental aid in this endeavor. In May 1990 – one month after the "dolphin safe" announcement – StarKist asked the US government to subsidize vessels which relocated and to negotiate the expansion of fishing terms and access in other tuna fisheries where regulation is less restrictive.[8]

This kind of behavior is nothing new, despite the hopes inspired by "dolphin safe." For decades, transnational tuna canning corporations such as Heinz/StarKist and large-scale commercial purse seine fleets have sought ever higher profits by playing the planet like a chessboard, continually moving operations to regions where natural resources, labor, and other expenses are cheapest and regulation most lax.[9]

II. *Technical Limits*

"For quite some time, StarKist...has consistently been at the forefront
in support of research related to dolphin and marine protection."
– *Anthony J.F. O'Reilly, Heinz Chairman, 1990*[10]

Heinz/StarKist has admitted the dangers of setting purse seine nets on dolphin. However, this did not prevent the company from distributing a brochure which argued that purse seine equipment and techniques were "perfected" to reduce dolphin deaths substantially.[11] Heinz/StarKist acknowledges that "no system is foolproof." But it attributes problems with purse seine nets and dolphin encirclement to factors such as weather conditions or skipper inexperience, rather than address the hazards inherent in the method itself.

StarKist asserts: "The goal must be no dolphin deaths and no dolphin injury."[12] If Heinz genuinely endorsed its subsidiary's commitment, one might reasonably expect the company to invest some of its considerable profits – derived in part from the sale of canned tuna – toward the development of sound alternatives for purse seine nets and the encirclement of dolphin. In 1990, Heinz/StarKist's competitor Bumble Bee Seafoods agreed to donate US$500,000 to an international research program for alternative fishing methods and equipment. But as of 1994 Heinz/StarKist had spent no money whatsoever on international programs to develop alternative technologies and practices.

III. *Environmental Limits*

Even if it were possible to create and manage a completely reliable program to protect dolphin during commercial tuna fishing operations across the globe and to monitor fishing vessels adequately, there would still be additional serious environmental problems.

Besides dolphin, modern commercial tuna fishing operations set nets on objects floating in the ocean such as logs under which not just tuna but many other aquatic species may congregate.[13] Once the nets are hauled in, the tuna are kept while the remaining ocean life, which can include other fish, sharks, marine turtles, sea birds, and dolphin, is discarded as "waste." Moreover, tuna gather around other species including whales and whale sharks, and purse seiner sets are frequently made around these creatures. Sets are also made on juvenile tunas, which can seriously endanger fish populations.

It is thus a mistake to believe – as Heinz/StarKist would have the public think – that "dolphin safe" alone actually means environmentally sound. As practiced by Heinz/StarKist, "dolphin safe" may in fact encourage the fishing industry to replace one destructive fishing practice (setting on dolphin) with other harmful fishing practices (setting on whales, setting on young fish, or setting on floating objects under which many different species gather). And given the nature of the commercial tuna industry and market, pressure exists to catch increasing amounts of fish. Already, the consequences of tuna overfishing are apparent. Nearly 20 years ago, for example, the breeding population of bluefin tuna in the Western Atlantic was ten times larger than it is today.[14]

But such pressure affects more than tuna. In recent decades, there has been a frightening decline in fish stocks globally, caused by the huge expansion of large-scale commercial fishing fleets and poor or non-existent management of the world's marine resources.[15] The UN Food and Agriculture Organization classifies 60 per cent of the fish types it monitors as fully exploited, overexploited, or depleted.[16]

IV. *Other Limits*

The activities of the commercial tuna industry adversely affect peoples of less-industrialized nations in several ways. The majority of tuna comes from tropical waters near many poor countries, where fish can be a far more significant part of people's diets than it is in most of the industrialized world.[17] Overfishing depletes these countries' food resources and harms the habitats necessary for local communities to survive. Because much tuna harvested in these waters goes to markets in the US, Western Europe, and Japan, commercial tuna operations do little to alleviate hunger in poor areas of the world.

In addition, the model established by US companies to develop the commercial fishing industry is now being copied by countries in Latin America and other less-industrialized regions, threatening to drive many small-scale fisherfolk out of business and adding to the problems of overfishing.[18]

Furthermore, the "dolphin safe" policies espoused by companies such as Heinz/StarKist have become an excuse by Western nations to boycott certain competing Latin American fleets and canneries.[19] Heinz/StarKist's refusal to help fund international initiatives to create and share safe fishing technologies with less-industrialized nations exacerbates this problem.

Heinz/StarKist's "Dolphin Safe" vs. A Precautionary Approach

> "As a company with consumers in more than 200 countries, we find the environmental issue to be irresistibly global. The great chain of procurement, processing, packaging, and purveyance is, at each link, related to the state of our planet's ecology. This gives us a direct stake in the outcome of the environmental debate that presently occupies the international agenda."
> – *From Heinz's Summer 1990 Quarterly Report* [20]

Heinz/StarKist claims to have a stake in global environmental issues. It connects that stake to the international scope of its business. It advertises its leadership in dolphin protection. For all these reasons, Heinz/StarKist should be a supporter of international programs to manage and control global fisheries and develop alternative technologies which ensure that dolphin, fish, and marine ecosystems are safe everywhere in the world. It should also be an advocate of global regulation, control, and monitoring to verify that safety. Heinz/StarKist is neither.

For too long, the tuna and other fishing industries have been given the benefit of the doubt about the possible consequences of their activities. It is time to adopt a genuinely precautionary approach to fishing, that is, when an activity could cause environmental damage, even if scientific evidence is uncertain, action should be taken to control the activity. If information to evaluate the impact of a type of fishing is insufficient, that activity should not be considered "safe."

A real commitment to environmentally sound tuna fishing practices demands more than advertising and new labels. A fishing industry genuinely committed to such practices would be properly managed, controlled, and monitored and would no longer deplete stocks of fish, disrupt the lives and livelihoods of poor communities, or catch and waste enormous quantities of marine species including, but not limited to, dolphin.

Endnotes

1. "Common Questions Concerning: 'Dolphin Safe' Tuna," in press release of StarKist Seafood Company, 12 April 1990.
2. Quoted in StarKist Seafood Company, Consumer Affairs Department, brochure, "Fishing the Eastern Tropical Pacific The Tuna/Dolphin Issue," 1989.
3. Estimate according to calculation of Greenpeace.
4. Quoted in StarKist Seafood Company, Consumer Affairs Department, brochure "StarKist Dolphin Safe,"
5. For more information see: Juan Carlos Arbex, "Pescadores Espanoles," Ministerio de Agricultura, Pesca y Alimintacion, Ruan S.A., Avenida de la Industria; 33 28100 Alcobendas (Madrid); Jacques Maigret, "Rapports Entre Les Cetaces et la Peche Thionere dans L'Atlantique Tropical Orien- tal," Imp. F. Paillart, Abbevilles No 5901, Juillet 1984. edit. 171; J. Levenez, A. Fontenau, R. Regalado, "Resultats D'une Enqute Sur L'Importance des Dauphins dans les Pecherie Thoniere FISM," SCRS/79/105; A.L. Coan & G.T. Sakagawa, "An Examination of Single Set Data for the US Tropical Tuna Purse Seine Fleet," ICCAT, 1982, v. XVIII, no.1; Apollo-Mary Elizbeth, Zapata Pathfinder, "Pacific Tuna Development Foundation," final report, Tuna Purse Seine Chapter to the Western Pacific, July-Nov. 1977; S. Leatherwood et al, "Observations of Cetaceans in the Northern Indian Ocean Sanctuary," Nov. 1980-May 1983; D.C. Simmons, Purse Seining Off Africa's Coast, Comm. Fish Review, March 1968; D. & M. Caldwell, "Porpoise Fisheries in the Southern Caribbean – Present Utilizations and Future Potentials," Proceedings of the 23rd Annual Session, Rosentiel School of Marine and Atmospheric Science, Willemstad, Curacao, Nov. 1970; A. Di Natale, Marine "Mammals Interactions in Scombridae Fisher- ies Activities: The Mediterranean Case," Aquastudio – Via Trapani, 6- 98121 Gessina,Italia. ICCAT 1990 Collected Volume XXXIII; and M. Louella & L. Dolar, 'Interaction Between Cetaceans and Fisheries in the Visayas, Philippines: A Preliminary Study," SC/090/629, Marine Labora- tory, Silliman University, Dumaguete City, 6200, Philippines. For an overview of the problems associated with commercial tuna fishing see: Greenpeace International, "In the Race for Tuna...Dolphins Aren't the Only Sacrifice The Impacts of Commercial Tuna Fishing on Oceans, Marine Life, and Human Communities," Amsterdam, 1993.
6. Inter-American Tropical Tuna Commission resolution establishing the Inter-Governmental Agreement (IGA), June 1992.

7. Private communication from Greenpeace International Ocean Ecosystems Campaign.
8. StarKist testimony before US Congress, 3 May 1990, quoted in Greenpeace's "Chronology related to H.J. Heinz Company and consolidated subsidiary StarKist Seafood Company," 19 August 1992.
9. Greenpeace, "Dolphins, Tuna and Free Trade: A Greenpeace Perspective," 1992, p. 2.
10. Quoted in Heinz's "Quarterly Report of Activities," Summer 1990, p. 3.
11. "Fishing the Eastern Tropical Pacific," op cit.
12. From "StarKist Dolphin Safe" brochure, op cit.
13. See sources from note 5.
14. Anne Swardson, "A loss that's deeper than the ocean – overharvesting is devastating the world's fish population," in *The Washington Post National Weekly Edition,* 24-20 October 1994.
15. Ibid.
16. Ibid.
17. Ibid.
18. Ibid.
19. For information on the situation between Mexico and the US see Kristin Dawkins, "NAFTA The New Rules of Corporate Conquest," Open Magazine Pamphlet Series, Westfield, New Jersey, 1993, pp. 10-11.
20. Remarks of Heinz CEO Anthony J. F. O' Reilly, in "Quarterly Report of Activities," op cit, p. 1.

Children of the Wilmington Montessori School participate in the 'RECYCLE DELAWARE' program by depositing their recyclables in igloos.

Source: BFI 1991 Annual Report.

"Our mission is to provide the highest quality waste collection, transportation, processing, disposal and related services to both public and private customers worldwide. We will carry out our mission efficiently, safely and in an environmentally responsible manner with respect for the role of government in protecting the public interest."
– *from BFI 1991 Annual Report*[1]

Greenwash Snapshot #20

BROWNING-FERRIS INDUSTRIES, INC

A case study in waste disposal.

Browning-Ferris Industries, Inc. (BFI)
Chief Executive Officer: William D. Ruckelshaus
Headquarters: PO Box 3151, 757 N. Eldridge Houston, Texas 77253 USA
Tel: 713-870-8100 Fax: 713-870-7844
Major business: solid waste disposal.

BFI operates in about 470 locations in North America and in 155 throughout Europe, Asia, Latin America, and the Pacific. BFI is a member of the WBCSD and a signer of the ICC Rotterdam Charter.

The BFI Mission

In its 1991 Annual Report, BFI spends over 20 pages trying to convince readers that it protects the environment, has the trust and goodwill of communities where it operates, and is responsive to government regulation. The discrepancy between BFI's public posture and its public record is great. Widely regarded as an infamous polluter, BFI has violated many environmental regulations; between 1979-88, the company's Environmental Compliance Listing cited 146 cases of waste disposal violations in 16 US states.[2] BFI also has a long history of antitrust law infringements.[3]

In 1988, BFI took steps to sanitize its image, hiring former US Environmental Protection Agency (EPA) Chief William "Mr. Clean" Ruckelshaus as CEO. Among other actions, the well-connected Ruckelshaus sold off BFI's notorious hazardous waste disposal affiliate, embraced recycling, and joined the BCSD.

Beyond its abysmal record, however, there is a contradiction inherent in BFI's mission. On the one hand, the company approves of waste minimization programs. At the same time, company executives acknowledge that such programs have contributed to a drop in revenue growth and profits while noting that BFI will remain dependent on its traditional business of landfills and its newer investments in incinerators for the foreseeable future.[4]

Landfills

The US Environmental Protection Agency asserted as early as 1981 that all landfills, even those with compacted clay liners or composite liners, will leak eventually.[5] A study for the EPA estimated that a ten acre landfill with the "best demonstrated available technology" for composite liners will have a leak rate of 73 to 3,650 gallons of fluid per year, or between 730 and 36,500 gallons over a decade. Design flaws or poor quality assurance would lead to a higher leak rate.[6]

Because certain toxics survive longer than any liner, landfills will pose a severe threat to underground water supplies for many decades to come.[7] But landfills remain the backbone of BFI's business. In 1991, solid waste collection and disposal in the US and Canada accounted for 88 per cent of the BFI's revenue, and this does not include operations in other countries or medical waste services. BFI dumped about 24 million tons of garbage into its 100 US landfills in 1991.[8]

Incinerators

In recent years, BFI has built two garbage incinerators in the US states of New York and New Jersey and begun building a third in Connecticut. As in so many communities in the US, the siting of these burners faced intense public opposition, and with good reason.

- Incinerators emit toxic chemicals into the air – even when operating at high efficiency. These include dioxins, furans, and thousands of other compounds.

- Incinerators do not destroy heavy metals but emit or concentrate them into residual ash, which must be landfilled.

- The main purpose of trash incinerators – to reduce waste volume – can be achieved without toxic emissions by waste reduction, reuse, recycling, and composting.

- Incinerators are expensive; BFI's New Jersey facility alone cost US$350 million.

- Despite the rhetoric of "waste-to-energy" plants, incineration actually generates very little energy.[9]

Recycling

In its 1991 Report, BFI proclaims: "Our goal, quite simply, is to be the premier recycling company in the country, and we believe we are well on our way." But the company still has a way to go; recycling services accounted for just three per cent of BFI's business in 1991.[10]

Moreover, what BFI considers recycling may not be environmentally beneficial. BFI is one of many companies which exports plastics for "recycling." Hong Kong, Indonesia, and the Philippines are the main recipients, and other countries in Africa, Latin America, Central and Eastern Europe, and the Caribbean have received US plastic scrap.[11] Some countries actually ban these waste imports, but underfunded customs infrastructures often fail to detect the violations. In 1991, BFI exported at least 895,000 pounds of plastic waste collected in the southeastern US to, among other places, Indonesia.[12] A BFI manager explained why, saying simply, "It's a third world nation."[13]

Much of the plastic scrap that BFI and other companies export may not even be recycled. A manager of a plastics recycling plant in Indonesia estimates that up to 40 per cent of the material is of low quality, contaminated, or otherwise unusable and thus ends up landfilled (see Solvay case study for more on plastics recycling).[14]

Solid Waste Disposal Problems

Across the United States, BFI has been embroiled in controversy surrounding its solid waste disposal practices. In 1988, for example, the
Massachusetts Department of Environmental Quality Engineering fined
BFI US$150,000 for consistently dumping more trash than was permit-
ted into its Randolph landfill. "You get fined for providing a public
service," complained BFI Vice President Peter Watson. A year earlier,
the company had been fined US$25,000 for improperly expanding the
Randolph dump.[15] In 1989, BFI was fined US$400,000 by the Louisiana
Department of Environmental Quality for "insanitary" practices at its
solid waste landfill near New Orleans. The company eventually paid
US$280,000.[16]

In 1988, potentially hazardous concentrations of methane gas were
discovered at an Eden Prairie, Minnesota landfill owned by a BFI
subsidiary and the landfill was closed. [17] In addition to the methane gas
emissions, according to Eden Prairie resident Jerri Collins, "The landfill
has polluted the ground water, the Minnesota River and the largest urban
wildlife center in the US. The site is on the Superfund list."[18] Despite the
site's dangers, BFI petitioned for two years to expand the dump. In the fall
of 1990, a study disclosed that new concentrations of methane gas had
spread from the landfill. BFI had withheld this study for months in
violation of the Minnesota Pollution Control Agency's (MPCA) rules.
City and state governmental representatives wrote to the MPCA Commissioner that "the landfill is dangerous and must not be reopened."[19]
Following the disclosure, BFI dropped its request to keep operating the
landfill.[20]

In 1991, a court ordered BFI to complete an environmental impact
report on its planned garbage dump in Azusa, California after the
company used questionable means to win permission for the dump
without such a report.[21] BFI invested US$100 million in the landfill and
hoped to earn US$800 million in disposal fees. CEO Ruckelshaus
appeared before the state's water board with a US$20.5 million offer to
help decontaminate groundwater in San Gabriel Valley. "When the going
got tough, they trotted him in there," a lawyer for the local water district
asserted, "It was an attempt to buy the permit." [22]

Other Legal Issues

BFI's record of antitrust violations began in 1972 when the Illinois Attorney General filed a suit alleging that a BFI subsidiary and 200 other garbage firms had conspired to fix prices and divide up customers. The companies agreed to pay a fine.[23] In 1983, BFI pleaded no contest to price fixing with competitors in Georgia.[24] In Ohio, BFI pleaded guilty to price fixing in 1987 and paid US$1 million in fines to the federal government and US$350,000 to the state.[25] That same year, a Vermont court ordered BFI to pay US$6.1 million in punitive damages for illegally undercutting prices in an attempt to drive a competitor out of business.[26] In 1989, the US Supreme Court upheld the US$6.1 million fine, which BFI argued was "grossly excessive."[27] And in Pennsylvania in 1990, BFI agreed to pay US$30.5 million for alleged nationwide price fixing with its leading competitor, Waste Management, Inc., which settled for US$19.5 million.[28]

> "There is absolutely no truth to any involvement by this company in crime of any kind."
> – *William Ruckelshaus, 1989* [29]

BFI and Hazardous Waste Disposal

In April 1990, BFI announced that it was withdrawing from the hazardous waste business. According to Ruckelshaus, hazardous waste "was 20 per cent of our revenue, none of our earnings, and most of our headaches."[30] Examples of why this was so abound.

In 1988, two hazardous waste subsidiaries of BFI, Cecos International and Chemical Services, Inc., agreed to pay US$2.5 million to settle federal and state lawsuits accusing them of improperly handling and disposing of toxic chemicals at a facility in Livingston, Louisiana. The EPA said it found over 1,700 violations of federal hazardous waste regulations.[31] In 1991, Cecos, which is not accepting waste at the site but still has cleanup operations, agreed to pay US$86,000 after the EPA charged that pollutant levels in water discharges had violated the company's permit.[32]

In 1990, Cecos and Chemical Services agreed to pay US$1.55 million to settle allegations that they had violated hazardous waste disposal regulations at a Willow Springs, Louisiana site. The EPA said that it had found 1,400 violations, including failure to maintain disposal pits adequately or to conduct inspections.[33] That same year, New York State authorities fined Cecos US$350,000 for violations at its toxic waste landfill in Niagara. It was the biggest fine ever imposed by the state for a hazardous waste dump. In addition, the state ordered a US$1 million remedial program to correct problems and denied a permit application to expand the site.[34] In 1991, the EPA said it was seeking US$14.2 million from Cecos for alleged illegal disposal of sludge contaminated with cancer-causing PCBs at the Niagara landfill.[35]

On 5 April 1990, Ohio Attorney General Anthony Celebrezze announced an agreement with Cecos to settle a case concerning disposal violations at the company's Williamsburg hazardous waste landfill. BFI agreed to plead guilty to a charge of illegally dumping toxic waste into a drinking water supply for 2,500 people and to pay US$3.5 million in civil fines and a US$25,000 criminal penalty to state and county officials. According to Celebrezze, "This case centered on one important principle – that corporations must be held accountable for their environmental practices."[36]

The same day as Celebrezze's announcement, BFI said it would pay US$295 million to close the Williamsburg landfill. The company also announced its withdrawal from the hazardous waste business.[37]

Endnotes

1.   Browning-Ferris Industries, Inc, 1991 Annual Report, p. 1.
2.   See Brian Lipset, "BFI: The Sludge of the Waste Industry," in *Multinational Monitor,* June 1990, pp. 26-28. For waste disposal violations see BFI Environmental *Compliance Listing,* part of a Pennsylvania required Form C Compliance History filed by Browning-Ferris Industries, Inc, from the files of Citizens Clearinghouse for Hazardous Wastes. CCHW's address is P.O. Box 6806 Falls Church, VA 22040 USA.
3.   See Lipset, op cit.
4.   Annual Report, pp. 32-33.

5. "EPA Says All Landfills Leak, Even Those Using Best Available Liners," in *Hazardous Waste News #37,* 10 August 1987.
6. In "Analyzing Why All Landfills Leak," in *Hazardous Waste News #116,* 14 February 1989.
7. "Decade-Old Study Revealed The Polluting Effects of Landfills," in *Hazardous Waste News #71,* 4 April 1988.
8. Annual Report, p. 6.
9. For information on the problems and hazards of incineration see: US Environmental Protection Agency, "Proposed Emission Guidelines for Municipal Waste Combusters," 40 CFR, Part 60, Federal Register, vol. 54, no. 243, 20 December 1989, pp. 52209-52304; Richard Denison & John Ruston, Environmental Defense Fund, "Recycling and Incineration," Island Press, 1990, pp. 178-196; Office of the Comptroller, City of New York, "Burn, Baby, Burn," January 1992, pp. 711-72; and Allan Hershkowitz, Natural Resources Defense Council, "Energy Savings and CO2 Reductions Resulting from Recycling," 1991 pamphlet.
10. Annual Report, p. 35.
11. From Greenpeace, Hazardous Exports Prevention Patrol Press Packet, Focus on Plastics and Plastic Waste Trade, 1 April 1992.
12. Ibid.
13. David Rogers, "Recyclers Making the Most of Our Trash," in *St. Petersburg Times,* 27 January 1991.
14. Greenpeace Press Release, 12 May 1992.
15. Stephen Shepherd, "BFI fined $150,000 by state," in *The Patriot Ledger,* 15 July 1988.
16. Lipset, op cit, p. 27.
17. "Eden Prairie, Minn., Elected Officials Ask MPCA Commissioner to Close Landfill," in PRNewswire, 15 August 1990.
18. Lipset, op cit, p. 27.
19. PRNewswire, 15 August 1990.
20. Jeff Bailey, "Trash Troubles Browning-Ferris Fails to Boost Its Business By Hiring 'Mr. Clean'," in *The Wall Street Journal,* 14 May 1991. Also "Water Agencies Hail Supreme Court's Azusa Landfill Ban," PRNewswire, 31 January 1991.
21. Bailey, op cit.
22. Bailey, op cit.
23. Lipset, op cit, p. 28.
24. Ibid.
25. Ibid.

26. Lipset, op cit, p. 28, and Bailey, op cit.

27. See BFI's response to allegation #3 of 20 December 1987 *Newsday* article, from files of Citizens Clearinghouse of Hazardous Wastes.

28. *The Wall Street Journal,* 8 November 1990.

29. Quoted in *Industry Week,* 17 July 1989.

30. Bailey, op cit.

31. "Two Browning Units Will Pay $2.5 Million In Settlement of Suits," in *The Wall Street Journal,* 15 August 1988.

32. Bob Anderson, "CECOS agrees to penalty of $86,000 for discharges," in *The Morning Advocate,* 19 December 1991.

33. "EPA Says Two Units of Browning-Ferris Fined $1.55 Million," in *The Wall Street Journal,* 9 March 1990.

34. Paul MacClennan, "Landfill addition blocked by DEC," in *The Buffalo News,* 14 March 1990.

35. "EPA Seeks $35.4 Million From GM, Two Landfills," in *The Wall Street Journal,* 19 March 1991.

36. CECOS Consent Decree, from release of Ohio Attorney General's office, 5 April 1990. See also Barnaby Feder, "Browning to Drop Waste Unit," in *The New York Times,* 6 April 1990 and "BFI Dumping Waste," AP, 6 April 1990.

37. Lipset, op cit, p. 26.

RESPONDING TO GREENWASH

Beneath the glossy public relations campaigns and the superficial environmental initiatives of the TNCs lie destructive production processes and products that are at the heart of the global environment and development crises. Neither corporate environmental departments nor green advertising can make a TNC whose lifeblood depends on toxic chemicals or nuclear reactors a friend of the environment. Despite the urgent need for globally binding agreements controlling the impact of TNCs on the environment and their role in development, the United Nations and other international organizations are all but ignoring the issue. Instead, corporations themselves and their political allies have set the terms of the debate.

Millions of people around the world, having borne the brunt of TNC depredations like those described in this book, understand that the corporate perspective on environmental and social issues is self-serving. Their experience demonstrates what common sense tells us: TNCs are not primarily interested in environmentally sound, sustainable development and cannot be relied upon to police themselves on environmental and development issues. No amount of consensus building or "conflict resolution" will erase this simple fact.

Control of TNC behavior must come from organized grassroots activism, participatory governmental processes, the force of local and national laws, and commitment to international agreements. In that context, corporate use of market mechanisms, "multi-stakeholder" processes, and sincere voluntary initiatives can be important; they cannot, however, replace regulatory control.

The first step in responding to greenwash is to reclaim the goals, values, and language of environmental movements. No one, whether advertising executive, magazine publisher, parent, or educator should allow the TNC vision of environmental protection and sustainable development to take over the hearts and minds of human beings. The environmental public relations blitz of TNCs should be analyzed, countered, and controlled. Governments, schools, churches, non-profit groups, and even the private sector itself can all play a role in this healthy and necessary skepticism of corporate environmentalism.

Close analysis of corporate environmentalism will in turn create space in which every individual, every community, and every region can develop hopes and goals for protection of nature and a clean environment. With those goals in mind, criteria can be developed against which TNC operations and investments should be measured.

Such comparisons will find that certain products and practices are simply beyond the pale, and must be phased out globally. Others, such as recycling, may be beneficial in some settings, but destructive in others. There is no economic activity which can automatically be considered environmentally sound regardless of where it is conducted. Still, we will all benefit if certain technologies, such as solar power, receive more encouragement than they currently do.

Measures of environmental soundness must come from those affected by economic activities, not just from the corporations themselves. These measures and criteria will naturally vary from country to country, community to community. The box on the following page describes one group's (Greenpeace's) criteria for Clean Production, which the organization believes all TNC activities should aim to meet.

From Greenwash to Green

Community and environmental organizations have long been active, often confrontational, in response to corporate destruction of the environment. This activism has resulted in a number of positive changes in corporate behavior, and has also led to many national and local regulations.

But in the age of unrestricted globalized capital flows, such regulations are not enough. As TNCs spread their influence, efforts by citizens groups, non-governmental organizations (NGOs), and others to globalize resistance to TNC power and environmental destruction are also underway, and must continue. After years of frustration in the attempt to draft even a simple Code of Conduct for TNCs, the task of confronting TNC power

What is Clean Production?

Greenpeace's Clean Production Definition and Criteria

Clean Production systems for food and manufactured products are:

✔ non-toxic;

✔ energy efficient;

✔ made using renewable materials which are routinely replenished and extracted in a manner that maintains the viability of the ecosystem and community from which they were taken, or;

✔ made from non-renewable materials previously extracted but able to reprocessed in an energy efficient and non-toxic manner.

Furthermore, the products are:

✔ durable and reusable;

✔ easy to dismantle, repair, and rebuild;

✔ minimally and appropriately packaged for distribution using reusable or recycled and recyclable materials.

Above all, Clean Production systems:

✔ are non-polluting throughout their entire life cycle;

✔ preserve diversity in nature and culture;

✔ support the ability of future generations to meet their needs.

By measuring our production systems against each of these criteria, we can strive for the goal of Clean Production rather than make regressive decisions and possibly more fatal mistakes.

Source: *Poland – The Green Tiger of Europe? Clean Production – The Only Way Forward*, Greenpeace International.

over ecological decisions affecting everyone on the planet remains one of the most important challenges, not only for the environmental movement, but for all social and political movements.

The real responses to corporate greenwash are the thousands of campaigns worldwide to defend jobs, land, culture, and the environment. Some of these campaigns have come forth with guiding principles of their own.[1] One example is an "alternative " treaty on TNCs, drafted by NGOs at the 1992 Rio Earth Summit. Such principles alone will not guarantee a future of environmentally sound, socially equitable development. Even if adopted, they too will be sometimes co-opted and corrupted by industry in another round of greenwashing. Implemented in a dynamic, democratic process, however, they CAN help combat greenwash and return the activities of TNCs under the control of the people whom they affect.

Some of these principles are outlined below.

I. Precautionary Action

The principle of precautionary action arose due to increasing recognition that past approaches to environmental protection against chemical hazards, based on the assumption that the environment has an "assimilative capacity" to receive and render harmless the vast quantity and variety of toxic pollutants, had failed. Prior to the 1980s, marine environmental policy in particular was based almost universally on assumptions of "assimilative capacity."[2] More and more, it is understood that this approach is not applicable, and indeed dangerous and irresponsible when applied to substances which are toxic and persistent.

Even the most sophisticated environmental impact assessment models, for example, contain substantial inherent uncertainty due to the overwhelming diversity and complexity of biological species, ecosystems, and chemical compounds entering the environment. What were once considered safe levels of particular inputs into the environment have subsequently been determined unsafe. The legacy of environmental degradation attests to this.

Responsibility to future generations suggests that all emissions of substances which are both toxic and persistent must be eliminated and prevented. It is irresponsible to permit emissions of these compounds which persist in the environment to degrade ecosystems and human health for years – in some cases hundreds and thousands of years. There is clearly no "assimilative capacity" of the environment for such substances.

Fortunately, a host of environmental fora have recognized the shortcomings – and dangers – of an environmental policy based on outdated "assimilative capacity" assumptions, and have adopted the principle of precautionary action. There are four components to this principle:

1. Shift in the burden of proof

Typically, those who engage in or propose an activity which risks harm to the environment take the position that others who question the activity must prove that it is harmful. As a general principle, such an approach is inadequate, because frequently it is only the proponent of the activity who is in a position to perform the necessary studies and assessments. It is especially inappropriate when the activity at issue involves toxic and persistent substances.

The precautionary approach shifts onto the proponent of an activity the burden of demonstrating that it is not likely to harm the environment or human health. This is analogous to food and drug regulations which adhere to the precautionary approach for a wide array of naturally and artificially produced products.

2. Prevention of contaminants entering the environment

The principle of precautionary action must reflect an approach based on prevention of contaminant inputs, rather than the common (but outdated) notion of attempted control of contaminants based on "assimilative capacity" assumptions. Precautionary action must not reflect a permissive approach based on "allowable emissions." It must be based on prevention and elimination of contaminants at the source. Zero input levels for designated substances should be a firm objective. To be truly precautionary, a definition should also state that the precautionary approach entailing prevention and elimination of inputs (zero inputs) should be applied to all persistent unnatural substances, as well as to all naturally occurring substances which are toxic and persistent.

3. Action before damage – before conclusive proof

The principle of precautionary action requires preventative action before waiting for conclusive scientific proof regarding the cause and effect relationship between contaminants and resulting ecosystem damage. Too often, such proof (when even attainable) comes after the fact and after the damage has already been done.

4. Implementation through Clean Production methods

Clean Production should be the goal of any new investments and development projects. Virtually all international fora which have addressed the implementation of the principle of precautionary action have appealed for eliminating and minimizing hazardous wastes and products through the application of Clean Production methods. Many of these fora recognize that it is essential to require a waste prevention audit of all individual plants and companies in order to (1) identify substances targeted for phase-out programs and (2) identify the corresponding Clean Production methods to achieve the phase-out.

Pollution prevention methods are one of the most important steps toward Clean Production. The traditional focus on disposal options and end-of-pipe measures should be replaced with a priority focus on substitution to eliminate the generation of hazardous wastes and products. The primary advantage of this approach is that it ensures that pollution will not be transferred to another pathway or environmental medium.

Pollution prevention methods are currently available in many industries. An essential component of their application is the use of mandatory environmental audits (see below) coupled with action programs with concrete phase-out targets tied to timelines. Although in its pure form Clean Production embraces concepts beyond pollution prevention, in practice the terms are sometimes used interchangeably.

II. Freedom of Information

There is a direct relation between citizen access to information and environmental quality. Individuals, community-organizations, and NGOs have often been the driving force for environmental reform, and require access to information for their efforts. The role of these groups is also vital in disseminating such information to the general public.

The Brundtland Report passed by the United Nations General Assembly in 1987 recognized this need and listed the following as a prerequisite for sustainable development: "a political system that secures effective citizen participation in decision making." The report further recommends that governments recognize:

> "the right of individuals to know and have access to current information on the state of the environment and natural resources, the right to be consulted and to participate in decision making on activities likely to have a significant effect upon the environment, and the right to legal remedies and redress for those whose health or environment has been or may be seriously affected."

Freedom of information includes:

1. Open public access to all government files and statistics, film, video, and computer information regarding the environment and natural resources, and copying rights to such information available at a reasonable rate;

2. Open access to all environmental audits and environmental impact assessments;

3. Open access to all waste streams, in all environmental media, for public, independent, sampling and analysis; and

4. Business confidentiality or "proprietary information" or "trade secrets" never being used as a rationale for denying information that is

relevant to assess the human health or environmental hazards of an activity.

Once a systematic and concerted transfer of knowledge is assured, there can be full and informed community involvement in the preparation of environmental impact assessments and decisions pertaining to project proposals. The "right to know" is an essential prerequisite to public participation and legal remedy.

III. Environmental Impact Assessments and Audits

Environmental Impact Assessments

An environmental impact assessment (EIA) should be undertaken by an independent, external consultant, with the full consultation of the public and regulatory bodies, prior to undertaking any new TNC projects. The EIA should list all raw materials, intermediates, products, and wastes, and should provide safety data for all substances used. In this way, the EIA would identify possible adverse environmental consequences prior to project approval, and would allow the consideration of alternatives which could adequately mitigate such consequences.

Where the investment entails the retrofitting of existing industrial plants with pollution control devices, the EIA should avoid the expensive "bandage" approach and implement Clean Production alternatives instead.

Goals of an EIA

The objective of an EIA is to ensure that the options being considered are environmentally sound and that any environmental consequences are recognized early in the project cycle and taken into account in project design. One result of the EIA could be that a project does not proceed.

The EIA should be carried out prior to undertaking any investment in order to:

1. Present predictions on environmental impact of proposed projects, with a view to identifying possible adverse environmental consequences;
2. Allow the consideration of alternatives that can adequately mitigate such adverse consequences; and

3. Allow the public – particularly those who will be affected by the project – and regulatory authorities an opportunity to comment on the proposed project.

The EIA should be an integral part of all investment projects. The same assessment, monitoring, and citizen participation requirements should be applied to all investments, whether involving agency-funded projects or privately funded joint ventures.

Elements of the EIA Process

1. The EIA report should include, at a minimum:

(a) Descriptions of the project and its surrounding environment;

(b) Analysis of its expected effects on the environment, not only in the construction and operational phase of the project, but also of the product (to be manufactured) itself, taking a cradle-to-grave approach;

(c) Justification for the project, including environmental, economic, and social considerations;

(d) Description of practical alternatives, including Clean Production, product substitution, and the no-action alternatives, and an explanation of why the project was chosen above all other alternatives;

(e) Analysis of measures proposed to mitigate expected adverse effects;

(f) Commitments to monitor and enforce any mitigation measures required;

(g) A brief, non-technical summary of the information provided under the above headings; and

(h) A statement of the extent to which the regulatory authorities and public bodies, including NGOs, in the country where the investment is to take place, were involved in planning the project, including a description of project modifications resulting from such consultation.

A discussion of alternatives is critically important because it sharpens analysis of the trade-offs between project benefits and environmental costs, and allows the decision maker to consider a range of cost-benefit configurations.

2. In order to allow a full assessment to be undertaken, further information which must be provided by the investor to the host government and the community impacted by the project proposal, must include the following:

(a) A copy of corporate guidelines for health and safety practices;

(b) A copy of relevant health and safety laws, regulations, and standards for all materials, processes, and products from the firm's home country and other countries of operation;

(c) Information on the history and performance of existing and closed plants, where similar processes and products are or were used, where the investor had full or partial ownership;

(d) A list and explanation of all fines, violations, criminal convictions, accidents, consent orders or decrees relating to the firm's operations or the marketing of the products;

(e) The results obtained from continuous pollution monitoring where similar plants are already operating;

(f) A list of all raw materials, intermediates, products, and wastes;

(g) Material safety data sheets for all substances used; and

(h) A list of all occupational and environmental health and safety standards, for materials and processes used and wastes generated.

3. The information provided as part of an EIA should be examined, or "audited," by an independent body as to its reliability. Before a decision is made to go forward with a project, government agencies and interested members of the public should be allowed an opportunity to comment on the EIA report.

Local government officials and interested parties should be consulted at every stage of project design, preparation, implementation, and monitoring, and in particular in the preparation of EIAs. This consultation is important in order to understand both the nature and extent of any social or environmental impact, and the acceptability of proposed mitigation measures. Public review, comment, and grievance procedures should be established. Confidentiality of information for any reason should never be used as a rationale for withholding information necessary to assess the hazards of an activity to the environment or to human health.

Enironmental Audits

An environmental audit is the equivalent of an assessment, but for already existing operations. It is a systematic examination of the interactions between any business operation and its surroundings. This includes: all emissions to air, land, and water; legal constraints; the effects on the neighboring community, landscape, and ecology; and the public's perception of the operating company.

Environmental audits should be undertaken at least once a year and should cover the enterprise's interaction with the environment, compliance with regulations, and a review of the firm's policy and management systems related to areas such as pollution discharge, choice of raw materials, waste management, and product planning.

Environmental audits may be used by government agencies as an enforcement tool. Environmental auditing does not stop at an assessment of compliance with regulations, nor is it a public relations exercise.

Rather, it is a strategic approach to the organization's activities, providing a unique opportunity for individual companies to start up a comprehensive environmental policy and strategy.

Goals of an Environmental Audit

The ultimate goal of an environmental audit should be the identification of ongoing adverse environmental impacts and the replacement of destructive processes with environmentally benign alternatives. This can be done by:

1. The mandatory quantitative and qualitative identification of all uses of toxic substances, of toxic processes, and of toxic wastes;

2. The prioritization of waste streams, particularly the halogenated hydrocarbons and those chemicals that are persistent, bioaccumulative, and toxic;

3. The identification of applicable Clean Production substitutes and pollution prevention methods in all industrial production processes, on an industrial sector basis; and

4. The development of a plan, with a firm timetable, for the reduction of the use of toxic substances and processes, and their substitution by Clean Production methods.

Scope of Environmental Audits

Environmental auditing should apply to all industrial activities and, in principle, to agriculture. It should include:

1. A copy of corporate guidelines for health and safety practices;

2. A copy of relevant health and safety laws, regulations, and standards for all materials, processes, and products from the firm's home country and other countries of operation;

3. Information on the history and performance of existing and closed plants, where similar processes and products are or were used, where the investor had full or partial ownership;

4. A list and explanation of all fines, violations, criminal convictions, accidents, consent orders, or decrees relating to the firm's operations or the marketing of the products;

5. The results obtained from continuous pollution monitoring;

6. A list of all raw materials, intermediates, products, and wastes;

7. Material safety data sheets for all substances used; and

8. A list of all occupational and environmental health and safety standards, for materials and processes used and wastes generated.

Freedom of Information

The results of an audit should be used in the preparation of an environmental statement with recommendations for improvement.

Both the audit and statement should be verified by an external independent auditor. After verification by an external independent auditor, this information should be accessible to:

1. The workers in the concerned industrial installations;

2. The communities impacted by the concerned industrial installation;

3. The shareholders of the companies concerned;

4. The general public insofar as they may be interested in the industrial activity concerned; and

5. Governmental regulatory bodies.

Such findings should never be subject to any clause of industrial confidentiality.

IV. Strict Civil Liability

If TNCs are truly committed to environmental protection and the communities which host their activities, they will be willing to enshrine that commitment in legally binding liability agreements and contractual clauses, enforceable in any country of operation.

Strict ("no fault") civil liability for personal injury/loss of life, property damage, and damage to the environment should be imposed on all TNCs, their affiliates, and any lending institutions involved, in order to hold them accountable for their decisions. By placing the responsibility for risks clearly with the corporation, it strengthens the "polluter pays" principle and at the same time, makes Clean Production more attractive and economically viable. Therefore, liability should be seen as a tool for seeking not only redress of environmental grievance, but as a powerful means of preventing environmental degradation. It is also a means of integrating traditionally excluded social costs, such as environmental degradation, into economic decisions.

The main provisions of an appropriate and adequate civil liability regime to achieve environmental protection and social justice include the following:

1. Strict 'no-fault' liability shall apply to all investors, as the ones creating the environmental risk, for damage whether caused by accident, negligence, or intentional, and whether incremental or sudden;

2. Compliance with regulations shall not be grounds to limit liability;

3. Strict liability shall cover both damage to the environment and harm to society at large, as well as any individual personal injury or property damage;

4. When two or more parties invest in a project, including the enterprise(s), and the lender if any, they shall all hold joint and separate liability;

5. The investor shall be liable for the acts of its agents and those of its sub-contractors;

6. Liability applies to damage caused at any stage of the project, from the construction phase, through operation, to the decommissioning of the plant, and any subsequent disposal;

7. No limitations, or ceilings, shall be placed on damage awards;

8. No time limitation shall apply to the application of the liability provisions. Any person or company causing damage to the environment shall be subject to the liability provisions regardless of when the damage occurred and when its source is determined. Such application shall also include persons or companies which complete, abandon, or otherwise cease operations which caused the damage.

9. Public authorities and public interest groups shall have the right to bring lawsuits against the investor(s) for damage to the environment, in order to obtain:

    (a) an injunction to prohibit or cease the act causing damage to the environment;

    (b) reimbursement of expenditure arising from measures to prevent damage to the environment;

    (c) the restoration of the environment to its state immediately prior to the occurrence of the damage to the environment;

    (d) reimbursement of expenditure resulting from other measures considered appropriate in the event that the sites affected cannot be restored at reasonable cost; and

(e) compensation for the damage;

10. Mandatory insurance or other financial security, such as bank guarantee, sufficient to cover any liability, shall be required;

11. The liability of the investor shall not be limited by any contractual provision purporting to limit or discharge his liability;

12. National and international contingency funds (based on the "polluter pays" principle) and plans shall be established to provide compensation not payable under the civil liability system, for example, in the event of insolvency.

The foregoing provisions should be adopted within national legislation, as well as agreed by all investors to be included in all contracts, business agreements, and joint ventures.

V. Public Participation and Training

Clean Production can be assisted by meaningful public participation in production decisions. Such public participation should include:

1. Public hearings on all project and investment decisions which may impact the environment;

2. Public right to participate and submit comments as part of the Environmental Impact Assessment and Environmental Audit procedures;

3. The right to public voting procedures such as initiatives and referenda regarding environmental issues;

4. The right to interview and question governmental officials involved in environmental decisions; and

5. The right to legal redress and remedies for environmental or personal damage.

Meaningful public participation in investment decisions is made possible by training. Local regulatory authorities can most effectively police and counteract dubious investment proposals and operations. In order to do this competently, however, the public, industry, and local enforcing authorities need training in:

1. Environmental Impact Assessment procedures

2. Environmental Audit procedures

3. Environmental law

4. Liability law

5. Clean Production Methods

Training programs must be independent of industry control to ensure against conflict of interest.

VI. Bans and Phase-Outs

As the Earth Summit recalled: "Gross chemical contamination, with grave consequences to human health, genetic structures and reproductive outcomes and the environment, has in recent times been continuing within some of the world's most important industrial areas." The United Nations Environment Programme (UNEP) negotiations on land-based sources of pollution have added that such contamination is not limited to industrial areas, but has spread to remote areas of the globe, such as the Arctic, via long range transport of persistent organic pollutants.

Scientific evidence of the last 20 years demonstrates that chemical phase-outs and bans are the most effective form of preventing environmental and health damage from toxic chemicals. While pollution control and waste treatment often merely shift contamination to another medium, or delay the entry of harmful substances into the environment, the removal of toxic inputs and products actually eliminates the hazard. This is especially true of persistent toxic chemicals, the hazards of which remain in the environment for years or decades. The lesson of the history of chemical management in the industrialized world is clear: pollution prevention and toxics elimination work; pollution control and risk management do not. Bans and phase-outs are the most effective form of pollution prevention possible.

Agenda 21 recommends that governments should undertake the "phasing out or banning of toxic chemicals that pose an unreasonable and otherwise unmanageable risk to the environment or human health and those that are toxic, persistent and bioaccumulative...." This is one of the most important recommendations of the Earth Summit.

The priority for bans and phase-outs should be given to organohalogens and other persistent toxic substances for which there is no safe level of environmental contamination or human exposure.

Even national bans are not enough, because overall production and contamination levels can stay high through export of the banned substance or through shifting production abroad. Therefore, the world community must:

- prohibit the export of substances which are banned domestically; and

- create a list of substances slated for global phase-out.

Endnotes

1. The principles outlined in this section are adapted from Iza Kruszewska, *Avoiding Western Mistakes: A Guide to Clean Investment in Eastern and Central Europe,* Greenpeace International, 1991, and from *Screening Foreign Investments: An Environmental Guide for Policy Makers and NGOs,* Third World Network, 1994.
2. For more information on this issue, see Joe Thornton, "Risk Assessment for Global Chemical Pollution? The Case for a Precautionary Policy on Chlorine Chemistry," paper presented at the Annual Meeting for the Advancement of Science, Atlanta, Georgia, 18 February 1995.